Digital Media and Society

Digital Media and Society

Transforming Economics, Politics and Social Practices

Andrew White
The University of Nottingham Ningbo China

First published 2014 by
PALGRAVE MACMILLAN

Palgrave Macmillan in the UK is an imprint of Macmillan Publishers Limited, registered in England, company number 785998, of Houndmills, Basingstoke, Hampshire RG21 6XS.

Palgrave Macmillan in the US is a division of St Martin's Press LLC, 175 Fifth Avenue, New York, NY 10010.

Palgrave Macmillan is the global academic imprint of the above companies and has companies and representatives throughout the world.

Palgrave® and Macmillan® are registered trademarks in the United States, the United Kingdom, Europe and other countries.

ISBN 978–1–137–39361–6 hardback
ISBN 978–1–137–39362–3 paperback

This book is printed on paper suitable for recycling and made from fully managed and sustained forest sources. Logging, pulping and manufacturing processes are expected to conform to the environmental regulations of the country of origin.

A catalogue record for this book is available from the British Library.

A catalog record for this book is available from the Library of Congress.

For Shasha & Jiajia

Contents

Tables

Preface and Acknowledgments

This book has long been in gestation, encompassing my research in digital media from almost a decade in academia as well as ideas that I picked up in various digitization projects that I worked on before that, the first of which began almost 15 years ago! This knowledge has become more structured in my own mind over the past two years during my teaching of a final year undergraduate module, *The New Media World*, at the University of Nottingham's China campus at Ningbo (UNNC). Originally developed with Rena Bivens for the first cohort of graduates at the university in 2007–2008, this module has also been taught jointly or solely by Ned Rossiter and Stephen Quinn. Being the sole convenor of this module since February 2012 has enabled me to develop its themes in ways in which will become familiar to those of you who continue reading past this point. While the content in this book is solely my own, the input of the aforementioned teachers, as well as all the students I have had the pleasure to teach on this module, has no doubt provided greater clarity to my thinking on these complex issues. I would also like to thank Chris Penfold and Felicity Plester at Palgrave Macmillan, both of whom showed great diligence and patience in supporting this project and answering my many queries. The reviewers and copy-editors, as well as those who generously provided endorsements, deserve my gratitude too.

Like the module, the book's structure is based on digital media's impact in three main areas: politics, the economy and social practices. Part I uses the 'public sphere' as a normative in order to explore the way in which digital media challenges existing conceptual models of the 'public' and the 'private' through analyses of the digitization of scholarly materials and of online identity. The final (third) chapter of Part I considers the effect of this shifting boundary between the public and the private spheres through the development of a theoretical framework for a digital-media-inflected politics in the liberal democratic state.

Part II comprises two chapters on the global digital economy, the first of which emphasizes the benefits that have resulted from the incorporation of digital media technologies in the general economy, as well as the growth of the 'creative industries'. The second chapter of Part II is more critical in tone, identifying the existing and potential problems, like the volatility of computer-based financial markets and the threat to job security that digital automation poses, which are consequences of the increasing digitization of the world economy.

Part III reflects upon the importance of the user in contemporary digital media research. The opening chapter of Part III attempts to construct a theory of user interaction with digital media, which moves beyond the simplistic instrumentalism/technological determinism binary. The complexity of theorizing this use is demonstrated in this chapter with a discussion of online reading practices. Chapters 7, 8 and 9 continue to develop the theme of 'use' through the analysis of new social movements, discussion of the practice of surveillance and an account of the various inventive ways in which citizens in developing nations engage with digital media.

Save for the acknowledgements above, the fear of inadvertently omitting someone important means that I will not list every single person from whom I have drawn inspiration in my almost 15 years of working and researching in the field of digital media! I would, however, like to thank the School of International Communications for giving me research leave to finish this book, and the librarians at UNNC who speedily procured reading material for me. I would like to thank all my family, especially Mum, Dad, Fi, Tegan and Callum for their love and support, and the Xu family for providing a home during the period when the bulk of this book was written. My greatest appreciation goes to Jiajia, who has had to bear my frequent absences during the writing of this book, even if I was often only in the next room! My absences have become particularly troublesome in the three-and-a-bit months since Jiajia brought our wonderful baby girl into this world. I feel great relief that I can now spend much more time with Jiajia and Shasha, both of whom I dedicate this book to.

Abbreviations and Acronyms

AP	Associated Press
ARC	Augmentation Research Group (Stanford Research Institute)
ARPA	Advanced Research Projects Agency (US Department of Defense)
ARPANET	Advanced Research Projects Agency Network (US Department of Defense)
CAT	Computerised Axial Tomography
CCTLD/S	Country Code Top-Level Domain/s
CEO	Chief Executive Officer
CIA	Central Intelligence Agency
DARPA	Defense [US Department of] Advanced Projects Agency
DCMS	Department for Culture, Media and Sport (UK)
DDoS	Distributed Denial of Service-attacks
DNS	Domain Name System
EU	European Union
Fortune 500	Annual list of top 500 US companies
G8	Group of Eight (France, USA, UK, Russia, Germany, Japan, Italy, Canada)
G20	Group of 20 (Argentina, Australia, Brazil, Canada, China, France, Germany, India, Indonesia, Italy, Japan, Mexico, Russia, Saudi Arabia, South Africa, South Korea, Turkey, UK, USA, European Union)
GCHQ	Government Communications Headquarters (UK)
GM	General Motors
ICANN	Internet Corporation for Assigned Names and Numbers
ICTs	Information and Communication Technologies
IGF	Internet Governance Forum

ISP	Internet Service Provider
IT	Information Technology
ITU	International Telecommunication Union
JISC	Joint Information Systems Committee
MacBride Report	International Commission for the Study of Communication Problems (1980)
MAI	Multi-Lateral Agreement on Investment
MIT	Massachusetts Institute of Technology
MP	Member of Parliament (UK)
M-PESA	Mobile-Money (Swahili)
NASDAQ	An American stock exchange
NGOs	Non-Governmental Organizations
NHS	(UK) National Health Service
NIESR	National (UK) Institute of Economic and Social Research
NSA	(US) National Security Agency
NWICO	New World Information and Communication Order
OCLC	Online Computer Center
OCR	Optical Character Recognition
OECD	The Organisation for Economic Co-Operation and Development
PageRank	Google's algorithm for ranking websites in its search engine
PDA	Personal Digital Assistant
PRISM	Planning tool for Resource Integration, Synchronization, and Management
RFID	Radio Frequency Identification (Microchips)
SMN	Social Media Network
S&P	Standard & Poor's
TCP/IP	Transmission Control Protocol/Internet Protocol
TIA	Total Information Awareness
UNCTAD	United Nations Conference on Trade and Development
UNDP	United Nations Development Programme
UNESCO	United Nations Educational, Scientific and Cultural Organization
USA PATRIOT Act	Uniting and Strengthening America by Providing Appropriate Tools Required to Intercept and Obstruct Terrorism Act 2001

WELL/Well	Whole Earth 'Lectronic Link
WIP	World Internet Project
WIPO	World Information Property Organization
WorldCat	OCLC (Online Computer Center)'s global library catalogue
WSIS	World Summit on the Information Society
WTO	World Trade Organization

Part I

Politics and Digital Media: The Impact of Digital Media on the Public and Private Spheres

Introduction to Part I

This part focuses on the impact of the growth of digital media on the political structures of modern societies. While most societies make a general distinction between public and private affairs, the countries in which the concepts of public and private spheres assume most importance are those which are fully democratic. Those countries with the longest democratic traditions – mainly, but not exclusively, in North America and Western Europe – have conceptualized their politics in relation to identifiably separate public and private spheres. In these countries, the public sphere sits between the state and the private sphere as a space in which people can discuss without hindrance the important social issues of the day. Theorists of the public sphere like Jurgen Habermas (1991/1962) have stressed the importance of the mass media as the primary form of communication within this public sphere. This section agrees with Habermas, to the extent that it focuses on the implications of the latest developments in media (or, more precisely, digital media) for the sense of the public and the private in modern societies.

But this part also highlights a flaw in the work of Habermas and other theorists of the public sphere, namely, the tendency largely to exclude discussions of institutions other than the mass media that shape the public sphere. These institutions include libraries, archives, museums and universities. If the media's central position in the theory of many public sphere scholars is due to its influence on political debate, then should not the role of the so-called memory institutions (libraries, archives, museums) and universities in constructing knowledge on the great issues of the day also be addressed? This section does this by considering the contemporary confluence of the media and these institutions, namely, the migration of content from the latter to the Internet.

Chapter 1 will consider the impact of this shift on the construction of knowledge in contemporary societies. It theorizes this in terms of information-gathering regimes, models which highlight the core methodologies we use when we are seeking knowledge. The implications of the shift of scholarly information from mainly publicly owned memory institutions and universities to mainly privately owned digital platforms are addressed here.

Chapter 2 focuses on identity through an exploration of the impact of digital media on the type of rational, civic-minded individual that is a crucial constituent of the democratic public sphere. The chapter outlines the various debates about online identity that have taken place since the 1990s and emphasizes the relationship between identity and knowledge construction.

The third chapter of the section theorizes the relationship between the public and private spheres in modern society in light of the changes described in the first two chapters. With a strong focus on the work of Habermas, as well as theorists of the online public sphere like Papacharissi (2010), the chapter will analyse to what extent our traditional conceptions of the public and private spheres have been challenged and whether or not this calls for a new form of politics.

1
From the Public to the Private: The Digitization of Scholarship

Introduction

Depending on your perspective, digital media threatens to either destroy or revolutionize millennia-old scholarly practices. The way in which we seek information online as a means of helping us to construct knowledge differs significantly from tried and trusted academic methods. This would not be significant if we did not use the Internet so much for this specific purpose, but, as will be outlined in more detail throughout this chapter, even those of us who carry out academic research are increasingly reliant on the Internet. For this reason, we need to understand both how we search for information online and the extent to which our online practices help or hinder us in our quest for knowledge.

Traditional forms of knowledge creation are based on structures and processes designed to lead the reader carefully along the path from the gathering of raw information to the development of understanding. These structures and processes have involved the categorizing of information (*classification*), testing its authenticity (*provenance*) and exposing us (*access*) to a plurality of it (*universality*). These core principles have underpinned scholarship and its attendant institutions like universities, libraries and archives for centuries. As this chapter will go on to illustrate, the rapid growth of digital media to its present pre-eminence as platform *par excellence* for the dissemination of information has called into question the validity of these long-held principles.

To some, that is to be celebrated. After all, in many ways, scholarship has become easier as we do not have to physically situate ourselves in the archive or the library. It is also easier to search for sources without using cumbersome catalogues. But as more educational resources have moved from public institutions to online platforms, largely in the private sector, how does this impact on the ways in which we develop our knowledge about the world around us? This chapter will attempt to provide answers to these questions, beginning at the place where modern institutional forms of knowledge building began.

The library at Alexandria

Private libraries had been in existence for centuries, even millennia, before Greek antiquity. Battles (2003: 25) reports that the first libraries of clay tablets appeared in Mesopotamia around 5,000 years ago and, by the seventh century BC, a library at Nineveh not only contained an impressive 25,000 tablets, but also was organized in a semi-systematic way. But it was the Greeks who, under Alexander the Great, built in Alexandria just over two millennia ago what is commonly believed to be the first attempt to construct an institution that could properly be regarded as a research library. Alexandria represented the first serious institutional attempt to put into practice the principles that govern academic research to this day. While, for various reasons ranging from elitism to low levels of literacy, one of those principles, access, was not at the forefront of the librarian's priorities, Alexandria did develop rudimentary forms of classification, provenance and universality. As Simon Goldhill elaborates:

> Without the practices of the library, we wouldn't have the university in the form we have it today, we wouldn't have the organization of knowledge we have today, we wouldn't have the whole institutions of scholarship that we recognize. And that seems to me to be the sort of legacy that is really profound.
>
> (Bragg *et al.* 2009)

For these reasons, Alexandria is as good a place as any to begin a discussion on traditional forms of scholarship.

The library at Alexandria and universality

In its attempt to secure as much Greek literature as it possibly could, as well as substantial collections in other languages, Alexandria can be regarded as the first viable attempt to establish a 'universal' library (Bragg *et al.* 2009; Cavallo and Chartier 1991: 10; Manguel 2008: 22,

24).[1] There has long been a strong association between the concept of universality and knowledge, which has provided the inspiration for the construction of libraries, archives and museums that attempt to collect 'everything' (White 2008). In the case of Alexandria, this was manifest in the library's housing of a community of scholars in its *Museon*, thus making a concrete link between universality of collection and knowledge creation. Despite – or perhaps because of – being destroyed after a few centuries, the idea that it encapsulated lived on in projects like the eighteenth-century *Encyclopedie*, Dewey's nineteenth-century decimal library classification system and, of course, contemporary national libraries and archives; all these projects and institutions are predicated on the Alexandrian concept of an inter-relationship between knowledge creation and universal access to documents. This inter-relationship will be explored further in a later discussion on digital media, before which there will be a discussion on other scholarly values which Alexandria popularized.

Classification/cataloguing

Gathering together under one roof as much information as possible is futile if it cannot be made accessible in a way that is convenient for those who seek to access it. This requires that the incoherent mass of information that users can access is given a structure or, to use terms germane to libraries and archives, classified or catalogued. It is believed that one of the librarians at Alexandria, Callimachus, created the world's first alphabetical catalogue, a 120-volume set which provided references of the library's most important Greek authors (Manguel 2008: 50; Polastron 2007: 15). As well as utilizing the alphabetic system which had been invented by Alexandria's first librarian, Callimachus also created tables based on different categories of knowledge (McNeely and Wolverton 2008: 20). This created not only the classification of different forms of knowledge, but also a canon of important authors and texts. While the lack of historical sources makes any judgement necessarily provisional in nature, the very fact that the Museon (museum) attached to the library attracted the region's finest scholars suggests that there must have been some way for them easily to retrieve written material.[2] This theory is bolstered by our knowledge that the catalogue was alphabetical within the categories, a system which McNeely and Wolverton (2008: 21) argue was successful in making 'books readily and rapidly accessible to roaming encyclopedic intellects [the scholars *in situ* at Alexandria]'.

Provenance

Another important element of the Alexandrian library which retains its significance to this day is provenance, the need both to identify the creator/author of an individual record and to establish that it is the original version. The seeming desire for the Ptolemies to possess original texts rather than copies appears to support Derrida's (1996: 91) contention that there has always been an obsession in western culture with 'origin', best illustrated in the archive. Both Manguel (2008: 24–25) and Battles (2003: 31) report that the Ptolemies often did not return the scrolls that they 'borrowed' for copying. This might have been unintentional, but it is certainly plausible to suggest that their desperation to retain the original documents was precisely because of the importance they attached to provenance. Supporting evidence for this stance comes from Simon Goldhill's (Bragg *et al.* 2009) reporting of the lengths to which the Ptolemies would go to secure original texts, sometimes to the extent of paying huge sums of money for 'borrowing' them – as above, he states that once secured, these texts often were not returned to their original owners; this is in addition to their practice of impounding books from ships that docked at Alexandria. And there was certainly an important practical reason for this: copies almost always contained many textual inaccuracies (McNeely and Wolverton 2008: 17).

The core elements of information-gathering at Alexandria

What emerges from this short discussion on the Alexandrian library is a general formula for information-gathering, much of which survives to the present day. These are:

1. Order/classification – cataloguing.
2. Provenance – collected original documents.
3. Access – limited to a small number of scholars.
4. Universal – tried to collect everything.
5. Public institutions hold and control most information.
6. Nation-building – Alexander the Great was able to spread Greek culture throughout the Middle East.

The last of these elements, nation-building, constitutes the modern imposition of a term on a historical epoch that did not have the same political construct, or at least not in the form that we understand it today. Nonetheless, its inclusion is useful in illustrating the extent to which libraries have also played a wider political role. In relation to this chapter, that political role is associated with the relationship between memory institutions and the state, particularly during the emergence of the European nation-state in the eighteenth century, to which we will now turn.

Francis Bacon, evidential scholarship and the emerging nation-state

In the modern era, the idea that there was an inter-relationship between scholarship, universality, classification and provenance received intellectual ballast from Enlightenment philosophies on knowledge construction. An acceptance of the supposition that there was a connection between universality and the attainment of knowledge led to great Enlightenment projects like Diderot's eighteenth-century *Encyclopédie* and the increasing proliferation in that same century of *bibliothèques*, or catalogues (White 2008: 114–115). This demonstrates not only the enduring legacy of the Alexandrian library, but also the influence of one of the most prominent philosophers of this or any other age, Francis Bacon.

Bacon is famous primarily for his invention of 'induction', the idea that scientific theories should be based on the observation of large amounts of data or of experiments. A fundamental requirement of Bacon's philosophy is that a 'great storehouse of facts should be accumulated' (Sargent 1999: xx). Thus can be discerned a link between Bacon's scientific method and the universal library:

Once gathered, this experience had to be compiled into organized national histories, that could be printed and distributed throughout the learned world and thus could foster communication and the free exchange of ideas and information. As early as his advice to Elizabeth I in the 1590s, he had been urging the establishment of institutions that would advance this goal, such as '*a most perfect and general library,*' *containing all 'books of worth' whether "ancient or modern, printed or manuscript, European or of other parts"; a botanical and zoological garden for the collection of all plants as well as rare beasts and birds; a museum collection of all things that had been produced 'by exquisite art or engine'; and a laboratory 'furnished with mills, instruments, furnaces and vessels' (vol. 8., pp. 334–335)* [my emphasis].
(Taken from Bacon's Book One, aphorisms, and cited in Sargent 1999: xx)

Universalist projects became *de rigueur* for the emerging European powers, as demonstrated by the archives which were constructed during this period: the House of Savoy archive in Turin in the early eighteenth century; Peter the Great's 1720 St. Petersburg archive; Maria Theresa of Vienna's 1749 archive; the establishment of princely and civic archives in Warsaw, Venice and Florence in the 1760s and 1770s; the creation of the French national archives in 1790; and the establishment of the UK Public Record Office (PRO) in 1838 (Steedman 2001: 69). And the methods of the historians working within these institutions were remarkably similar to Bacon's notion of induction, where the evidence was believed to speak for itself. We see this in the figure of influential twentieth-century British archivist Hilary Jenkinson who, like Bacon, believed that the hypothesis should follow, rather than precede, the evidence (Gilliland-Swetland 2000: 12).

This type of evidential scholarship was based on principles that have altered little since Alexandria: namely universality, provenance and classification. The same could be said for librarianship as well, which, in the nineteenth century became more systematic in its acquisition and storing of books, especially after Melville Dewey's invention of a standardized decimal form of classification. Universality was promoted, like it was at Alexandria, through trying to collect virtually everything – in the UK and Ireland, for instance, there are six legal deposit libraries to which all publishers and/or authors in those territories must send two copies of their books.

As people became more literate throughout the nineteenth century and libraries and archives became, in theory at least, more accessible, these institutions had growing political influence. This role has been identified by McNeely and Wolverton (2008: 165) in the use, by nineteenth-century nationalists, of public education in an attempt to unify European societies which were riven with ethnic, religious and class tensions. Similarly, the UK's Public Library Bill of 1850 was underpinned by a utilitarian philosophy which supposed that giving people greater access to information would make them more disposed to 'reason' (Battles 2003: 137).

The core elements of information-gathering in the modern, democratic nation-state

Let us remind ourselves of the earlier general formula for information-gathering at Alexandria and compare it to that in the modern, democratic nation-state.

Alexandria

1. Order/classification – cataloguing.
2. Provenance – collected original documents.
3. *Access – limited to a small number of scholars.*
4. Universal – tried to collect everything.
5. Public institutions hold and control most information.
6. Nation-building – Alexander the Great was able to spread Greek culture throughout the Middle East.

Modern, democratic nation-state

1. Order/classification – cataloguing.
2. Provenance – collected original documents.
3. *Access – public libraries, museums and archives widened access.*
4. Universal – tried to collect everything.
5. Public institutions hold and control most information.
6. Nation-building – information-gathering institutions linked to power and the nation-state.

The only major difference between the two is that of access, which, as a result of greater levels of literacy, better modes of transport and a democratic impulse to share knowledge as widely as possible, gave citizens of nineteenth-century Europe much better opportunities than those in Egypt two millennia ago. That this is the only key difference illustrates the enduring legacy of Alexandria's values. Similarly, the nineteenth-century typology above also accurately represents the contemporary situation today, or at least until the recent exponential growth of digital media. This last point alludes to the argument of many that these information-gathering principles are no longer relevant to our modern informational environment, a debate to which this chapter will now turn.

The existential threat to the library and the archive

Those who first visit the impressive-looking building in downtown Washington, DC which houses the USA's national archives and records might be surprised when they discover that the National Archives and Records Administration (NARA) is dwarfed by its over-spill building in Maryland. Similar tales can be told about national archives and libraries in other countries, and are an enduring reminder of the capacity of the universal library to confound those who try to build it. The lack of affordable space in many of the world's capital cities, a problem made more acute by the strain on the public purse engendered by the global financial crisis, imperils the continued, seemingly unlimited, growth of national libraries and archives.

There is another threat to the national library and archive which is more existential in nature. These national institutions have been so successful in developing civic consciousness among their citizens that they are seen to embody the values of their nation. The logic of this is that their destruction will not only result in physical loss but will also threaten those very values that they are seen to embody. And such is the identification of the nation-state with these values – national archives and libraries help to shape public consciousness which both reflect and propagate their own nation's values – that in war-time the destruction of archives and libraries can cause considerable loss of morale among the citizens of the nations to which they belong. While the wilful destruction of archives and libraries can be traced all the way back to ancient Alexandria and perhaps beyond, the sophisticated technologies that modern armies have at their disposal means that this can be carried out in a much more efficient and systematic manner. During the Bosnian conflict in the 1990s, for instance, the Serbs targeted a number of cultural institutions, reaching an apotheosis with the destruction of most of the 1.5 million volumes in the National and University Library of Bosnia (Battles 2003: 188). According to András Riedlmayer, there was an, albeit twisted, rationale to this destruction:

> Throughout Bosnia, libraries, archives, museums and cultural institutions have been targeted for destruction, in an attempt to eliminate the material evidence – books, documents and works of art – that could remind future generations that people of different ethnic and religious traditions once shared a common heritage....
> (Riedlmayer, A. [full reference not given by Battles] cited in Battles 2003: 188)

And, in a further twist, the person who signed the directive ordering General Ratko Mladic to shell the Vijecnica neighbourhood within which the National and University Library of Bosnia stood was Nikola Koljecvic, a former Shakespearian scholar who had often patronized the institution during the cosmopolitan tranquillity of pre-war Sarajevo (Battles 2003: 186–187). As in other such incidents throughout history, this was not merely a by-product of war: perhaps better than anyone else in Bosnia, Koljevic realized the importance of the role of culture in conflict. As Battles (2003: 156) points out, it is likely that the actualization of the fear of loss has had a profound influence:

> It may not be too much to say that the sudden disembodiment of the book in the late twentieth century – as text disappeared first into the grainy obfuscations of microfilm and eventually into the pixelated [sic] ether of the Internet – began with crude renewals of violence against the book in the First and Second World Wars.

Alluded to above is one of the methods employed by archivists and librarians in response to the potential threats posed by diminishing space or destruction in war-time, namely micro-filming. Novelist Nicholson Baker's (1992) polemic on micro-filming practices in British and American libraries illustrates the unsuitability of this method of preservation. Baker details a litany of destruction of original newspapers, periodicals and books in some of the world's most prestigious public and university libraries in the process of their micro-filming. The Library of Congress's chief of photo-duplication in the 1970s, when it was the midst of destroying thousands of newspapers, Charles La Hood, wrote that: 'Microfilming came at a propitious time, as the Library of Congress was experiencing an acute space problem in its newspaper collection' (cited in Baker 1992: 35). This lack of space is, of course, not absolute, but an acknowledgement that the institution was either unwilling or unable to buy additional storage space, as Baker (1992: 36) would have liked it and other institutions to do.

Baker's polemic, though, has been challenged most prominently by Richard Cox (1992), whose critique of it touches on some of the issues that animate discussion of the purported impact of digital media on knowledge and society. This includes associating Baker's stance with the existential fear of losing information, as articulated by many western writers – perhaps most eloquently expressed in Ray Bradbury's novel *Fahrenheit 451* – and with the long-held dream to construct the universal library (Cox 1992: 11–14). He uses a quotation from a review by Julian Dibbell in the *Village Voice Literary Supplement* to emboss his point:

> Bush's [Vannevar] fantasy is no crazier than Baker's. In essence, both of them dream of having access to all the information ever published, and it drives them nuts to see a single scrap of it fall through the cracks. But the world and all the order in it are always slipping through the cracks, and the failure to reconcile oneself to that is,

> among other things, as good a definition of obsessive compulsion as any.
>
> (Dibbell 2001; cited in Cox 1992: 124)

While not commenting on the cost of Baker's proposal that public libraries and archives build warehouses to store material rather than micro-film it, Cox (1992: 43) argues that this would make collections less accessible to the public.

But the debate about micro-filming is now largely academic, as digitization has become the primary method for the reproduction of original scholarly sources. The advances in computing technology and storage capacity over the two decades since Cox and Baker's books were published would appear to supersede concerns about a trade-off between storage and access. In the world of paper archives and libraries, the increase in amount of material, entailing as it does the building of storage space on remote sites where property is cheaper, decreases the quality of access. In addition to providing an elusive target for those with baleful intentions, the virtualization of the archive and library offers a platform for a seemingly unlimited amount of scholarly material, as well as the capacity, through the operation of highly sophisticated search engines, to deliver any digitized document to the user's PC within seconds. This is an obvious lure for those of us who spend inordinate amounts of time both visiting different research institutions and waiting for material to be delivered from over-spill stores *within* institutions (in the British Library, an increasing number, possibly a majority, of books take two days to arrive at a reader's desk after a long journey from Boston Spa to London). Electronic access is purportedly egalitarian in its exposure of these collections to a much wider audience than those who have the time, money and inclination to travel to the best archives and research libraries (White 2011: 317–318). Accordingly, a number of great public institutions like the USA's Library of Congress and the British Library have long-running digitization projects, but their efforts are dwarfed by the search engine giant, Google, especially since the onset of the global financial crisis (White 2011).

The Google Books project

Since Google announced in December 2004 its agreements with some of the world's leading research libraries, it has made steady progress towards its goal of digitizing every single one of the 32 million books in the WorldCat (Stross 2008: 98, 107–108; Vise 2006: 238). The project is an extension of the corporation's belief that gathering together as much information as possible and making it easily retrievable will enhance humans' capacity for knowledge construction:

> It [Google] seek to develop 'the *perfect* search engine', which it defines as something that 'understands *exactly* what you mean and gives you back *exactly* what you want' . . . .
>
> In a 2004 interview with *Newsweek*, Brin [Sergey] said: 'Certainly if you had all the world's information directly attached to your brain, or an artificial brain that was smarter than your brain, you'd be better off.' [emphasis in the original]
>
> (Carr 2008: 4)

Explicit in Google's mission is the creation of an information-gathering regime that will not only supersede this author's schema for information-gathering in the modern, democratic nation-state, but will continue to evolve until perfection in information retrieval is attained. Like much that pervades digital media, this is a revolutionary approach in that it abandons the careful construction, over a period of 2,000 years, of an information-gathering regime that differed little from that during the time of the Ptolemies, for a model which is perpetually re-calibrated. Given the number of books that Google has already digitized – Auletta (2010: 257–258) claims that this was seven million by October 2008 – it seems that the universal library (of books) is within reach. While this is potentially a boon for researchers and the general public, it will profoundly alter existing information-gathering models.

But we should never lose sight of the metaphysical rationale for the building of these huge electronic libraries. Manguel's (2008) view that the desire for the universal library appeals to our need to establish a sense of order in a complicated world echoes Google's proclamation that its massive digitization project is not merely to make these books 'universally accessible and useful' but a more ambitious plan to '*orga-nize* [my emphasis] the world's information' (cited in Appleyard 2007). These large-scale digitization projects should not, then, be viewed only as a means of improving access to existing archives and library collections, but also as a re-organization of the information contained within them. The question that needs to be asked in relation to this is: what implications does this have for the type of evidential scholarship based on classification and provenance that has been the basis of knowledge construction for centuries? This question will be answered through a consideration of the extent to which the scholarly methods outlined earlier are applicable to our digital media ecosystem.

Information-gathering in Googleverse

Before considering the way in which the virtualization of so much of our scholarly heritage will affect how we search for information, let us remind ourselves of the existing typology:

Modern, democratic nation-state

1. Order/classification – cataloguing.
2. Provenance – collected original documents.
3. *Access – public libraries, museums and archives widened access.*
4. Universal – tried to collect everything.
5. Public institutions hold and control most information.
6. Nation-building – information-gathering institutions linked to power and the nation-state.

The sixth element, nation-building, can be set aside for now, and can be more fully considered in Chapter 3, on the theorization of politics in our digital age. How are the other five elements to be conceptualized in societies where the search engine is often the first port-of-call when we look for information?

It is easy to exaggerate the potency of the search engine, especially in the light of the findings of Head and Eisenberg's (2009) study of 2,318 USA college students that most of them referred first to course readings when they were writing assignments. However, those same figures show that around 96 per cent also used Google and 85 per cent used Wikipedia for help with their course assignments; the figure for course readings was 97 per cent (Head and Eisenberg 2009: 18). For information-seeking that is not related to their courses, students were most reliant on Google (around 98.5 per cent) and Wikipedia (around 90 per cent). A previous study by Head (2007) demonstrated that, even for research for course assignments, students were most likely (47 per cent) to go to the World Wide Web first. Other studies by the OCLC (Online Computer Center) in 2006 (cited in Rowlands *et al.* (2008: 292)) and Van Scoyoc and Cason (2006) report similar findings, with the former observing that 89 per cent of college students used commercial search engines when they began their research; and only 2 per cent started with a library website. In all these studies, the assistance of both library websites and librarians themselves was low on the list of students' priorities.

Thus the search engine's epistemological break with traditional models of information-gathering cannot be ignored when we are discussing knowledge construction in a digital age.

1. *Classification and the mathematical algorithm*

As stated throughout, even though electronic catalogues have been in use for decades in libraries and archives, the concept of classification has retained its value within those institutions. This is partly because it helps us to navigate our way around the increasing proliferation of information that digital media has ushered in. The need to manage large surges in information is not new; Postman has reported that there was a significant increase in the number of schools in England from the late fifteenth to early seventeenth centuries as a response to Gutenberg's print revolution (1993: 62–63). But earlier forms of information management have taken place within the broad information-gathering framework referenced throughout this chapter. Thus the search engine is so radical not only because of its technical capacity but for its rupture of existing information-gathering protocols.

And how is the search for information facilitated by search engines if there is no discernible method of classification? The short answer is that search engines use sophisticated mathematical algorithms to deliver to the reader the information most relevant to his/her search. Unlike traditional classification systems, these algorithms are secret and are being constantly re-calibrated. Google is thus the most efficient search engine because its engineers have designed a better algorithm than its competitors. The lack of obvious structure runs contrary to the type of evidential scholarship that has been the mainstay of the best models of education for centuries. There are, though, those who believe that we should celebrate the supersession of formal classification.

David Weinberger's book, *Everything is Miscellaneous,* is a paean to the wonders of a classification-free digital world (2007). Weinberger's basic argument is almost technologically determinist in its assertion that the rules of the so-called world of atoms do not apply to digital information. In relation to scholarly information, the tagging of each digital file with as much metadata as possible supersedes, according to D. Weinberger (2007: 17–23), the need for classification. This will enable researchers to access the most relevant information without the intercession of a scholarly guide. This belief is partly based on the idea that digital media has caused the death of distance, not only bringing those far-flung archives and libraries nearer to us through the act of digitization, but information generally, which is now at our 'fingertips' and hence does not need require a mediator (in the form of classification) in the same way that it did in the past (Friedman 2006: 176–185). To paraphrase Sergey Brin's words above, who needs mediation when 'all the world's information [can be] directly attached to your brain' (Carr 2008: 4)? And in an era where many of us are bypassing traditional library and archival classification systems and finding rich resources online, then there is some merit in this argument. But there are two major flaws in it too: the argument that there is no structure in the search engine is problematic, as is the belief that structure no longer matters.

Classification in academia has always been a problematic concept, as seen in Petrucci's (1991) argument that creating a canon of the best authors in any given discipline is a project designed to sustain governing ideologies. In this sense, the creation of reading lists and the classification of material in the library and the archive could be read in the same way; after all, national memory institutions are designed to sustain *national* memory perhaps even more so than general knowledge. These types of postmodernist critiques of academic classification are not new, but have been given greater potency by the advent of technologies that provide a viable practical alternative to formal classification schemes. But the belief that the mathematical algorithms that drive search engines are random or neutral is fundamentally flawed. Google's PageRank technology bears similarities with academic peer-reviewing in its ranking of websites. The algorithm elevates those websites that are physically linked to larger numbers of other websites, with links to the higher-ranking websites giving additional privileges. Gleick (2011) contrasts this with methodologies of earlier search engines which ranked websites purely quantitatively, emphasized by an anecdote about how the Oregon Center for Optics appeared as the first result of an *Altavista* search for 'university' simply because that word appeared many times in a headline about the Center.

One of the, no doubt unintended, effects of the PageRank methodology is that it reinforces rather than challenges existing hierarchies of information, a phenomenon O'Neil (2009: 58) likens to the 'Matthew Effect' in academia, whereby established researchers are much more likely to gain citations than their less experienced colleagues. Also, though their ideology is global, search engines are biased towards certain regions and languages. The most commercially successful search engines are American and there is a bias towards English language sources generally and American websites specifically in their rankings; this is partly as a result of their longer vintage (Halavais 2009: 89–90). In a widely publicized critique of Google, the then French national librarian, Jean-Noel Jeanneney, was concerned, among other things, about the search engine's bias towards English language sources even on subjects where sources in another language would be more appropriate. From his own country's perspective, there was a danger that people would be referred primarily to English translations of the work of France's greatest novelists or English language versions of significant events in its history (Jeanneney 2007: 42–43). The French government tried to counteract these tendencies through the development of a European search engine during the middle of the first decade of the millennium, but this – Quaero – petered out when Germany withdrew its support in 2007 (Doueihi 2011: 166–167; Vaidhyanathan 2011: 25). Despite this, Vaidhyanathan (2011: 138–139) reports that Google is increasingly tailoring searches based on location of the user. While this might be an efficient strategy for locating your nearest pizza parlour, searches that pander to the user's particularities are probably not the best way of developing broad-based intellectual knowledge.

2. *Reputation-building rather than provenance*

The tendency of techno-utopians to argue against traditional forms of classification while ignoring the way in which the Internet classifies information is repeated in relation to provenance. The students who, in the empirical studies earlier in this section, spurned the authority of the librarian for the supposed efficiency of the commercial engine can justify their actions by invoking a whole class of Internet theorists, social commentators and commercial corporations who believe that challenging the experts is the intrinsic duty of Internet users (Weinberger, D. 2007; Friedman 2006; Brin, cited in Carr 2008: 4). But those of us who teach in universities know that encouraging students, or anyone else for that matter, to launch critiques of established theories and concepts without reading authoritative texts is not good pedagogical practice. My impression is that, by and large, students are aware of this too. At the same time, the convenience of Internet searching makes it an attractive option that they are not going to forgo in their academic studies anytime soon.

There are some, especially the commentators mentioned in the last paragraph, who believe that the Internet represents a democratization of information and therefore is a welcome departure from authority-based knowledge development (Friedman 2006: 176–185). But the need to sift credible sources of information from the not-so-credible is accepted even by some of the most enthusiastic advocates of user generated content (Gillmor 2010). Provenance in the library and archive is partly based on the location of the book, document or record within a wider structure, be it a series of records or a canon of literature. The Internet is not structured in that way, so how does provenance operate within it?

The first approach is through the establishment of reputation. Online reputation is determined mainly quantitatively. This is the core of Google's PageRank which, while giving additional credence to websites linked to their highly ranked peers, ranks websites by the number of times they appear as links in others. The success of this method is illustrated by Vaidhyanathan's (2011: 59) review of some empirical studies which demonstrate that users exhibit a 'trust bias' in relation to Google's ranking. There are other websites, like Reddit, Digg and Del.icio.us, that are devoted to ranking reputation through the use of 'folksonomies': a method which allows users to tag those websites or sources of which they most approve (O'Neil 2009: 49–50). Some of these methods are taken from commercial websites, most notably Amazon's ratings of books and the reliability of second-hand booksellers (O'Neil 2009: 50). This has led to the championing of the 'wisdom of crowds' or the 'hive mind', a form of collective intelligence that the Internet can easily facilitate (Leadbetter 2008; Shirky 2009; Surowiecki 2005). The belief that the hive mind is superior to individual experts is manifest in projects like Wikipedia. But this approach dangerously conflates popularity with authority, risking in its extreme form, to use the delicious phrase of Jaron Lanier (2011: 79), 'digital Maoism'.

Like the dismissal of academic classification, the casual suppression of the principle of provenance in the online world does not actually push it to the margins but encourages it to mutate into a more virulent form. Thus, rather than exposing us to a greater plurality of information, all too often those with the greatest reputation online are interchangeable with popular figures offline. In the London *Independent* newspaper's 100 most influential 'Twitters' in 2012, included in the top ten were a footballer, a celebrity chef, a DJ and a famous illusionist, as well as four comedians/actors (Burrell 2012). This illustrates the populism, rather than pluralism, of the so-called 'hive mind' and much of what passes for information on the Internet today.

3. *Greater access?*

The Internet continues to give ever greater access to its content. While many websites are still censored in China, the world's most populous nation now has an estimated 590.6 million Internet users, representing

44.1 per cent of the population (Pew Research Center 2013a). On the African continent, where access to technology has traditionally lagged behind other regions, the recent rapid take-up of mobile phones and wifi technology has greatly improved access (see Chapter 9). In the developed world, the migration through digitization of scholarly materials to online platforms continues, seemingly unabated. While some people continue to have much greater access to the Internet than others, the global diffusion of digital media technologies seems to be closing this gap rapidly (Friedman 2006). But even this rosy scenario contains some caveats.

Let me explain this by reflecting on my own access to Internet content as I type these sentences on my university PC in China. In many ways I am information-poverished, as my access to websites like Google is severely restricted or blocked. I could, of course, buy a VPN to deal with this problem, but this involves a fee for what ideally should be free, could make me more vulnerable to viruses and advertising, and does not always work properly. Sadly, this tale of patchy coverage and increasing potential financial cost of surfing is all too familiar in an Internet where government restrictions are increasing in many places and where commercialization is taking a firmer hold. However, through my university, I have greater electronic access to articles from some of the world's leading educational journals and databases than do the vast majority of my fellow citizens. I can also access the *New York Review of Books* on my e-reader (which requires a small subscription each month) but not the London *Times* (which also requires a small subscription that I do not want to pay). My anecdote highlights the main difference between access in China and the Anglophone world.

In the Anglophone world, intellectual property serves as the most important gate-keeper to online content. Were I not a member of staff at a prestigious university which is prepared to pay the expensive institutional subscriptions for many of the leading journals in my own discipline, it would be difficult for me to have anything more than limited access to the journal articles of the main educational publishers. As my example of the London *Times* shows, an increasing number of the world's most prestigious newspapers are also retreating behind online paywalls. Access to copyrighted material is not so much of a problem (again, depending on your perspective) in China – where Montgomery (2010: 108) estimates that up to 90 per cent of film and music that is watched and listened to in China is pirated – but random blocking of access to content that does not have copyright restrictions clearly is. What we can conclude from all this is that, while the Internet is providing ever greater access, it is not clear what access models will dominate in the future. The Google Books project illustrates the folly of trying to predict which model will prevail. The project has slowed down after a number of legal challenges by publishers and at the time of writing it is not clear whether all these books, which the corporation has digitized will be accessible to the general public in the future. There is no guarantee that Google will even exist in a few decades' time; if it does, it is not beyond the realm of possibility that it might decide that providing free access to its digitized books is no longer a priority or economically viable (Vaidhyanathan 2011: 165, 202). With some European politicians, including the British Prime Minister David Cameron, thinking aloud that suspected rioters should be banned from using social media, we should not assume that the Internet is on an ever upward curve of greater access (Halliday 2011). What can be stated with more conviction, though, is that our present generation lives in a much richer information environment that at any point in history.

4. *OCR'ing technology, transcription and universality*

It is surely safe to assume that greater access has also been mirrored by a serious move toward the once seemingly impossible goal of the universal library. Notwithstanding its legal problems, Google's plan to digitize every single book in the WorldCat could be completed in this decade. This project is supplemented by initiatives to digitize special collections in libraries, archives and universities throughout the world (White 2011: 321). Does this mean that the universal (digital) library is attainable? The short answer is: not in the near to medium future; and a longer explanation follows.

What is good scholarly practice, namely the digital copy's fidelity to the original, is good for the universal library too. This is because inaccuracy in copying is not only detrimental to the quality of individual scholarly resources, but also reduces a digital library's coverage, in other words, makes it fall short of universalism. Every time text is omitted or obfuscated in the journey from the printed page to the electronic file, or metadata is not recorded, there is a reduction in the amount of information in the universal library. When these omissions and inaccuracies reach a certain level (Library of Congress (2013), guidelines for its digitization programmes stipulate that these should occur in no more than 0.05 per cent of total characters, or one in 2,000), then the universality of the library is called into question. Only a very few of the accounts or critiques of Google Books discuss the actual quality of its digital copies. Of those that do, Jones (2010) and Duguid (2007) identified an alarmingly high number of images that it would not be possible to convert into machine-readable text, while a more positive account by James (2010) identified errors in less than 1 per cent of the pages he sampled. It might be possible to 'clean up' the text by human hand, either individually or by crowd sourcing.[3] But cleaning up millions of words is time-consuming and expensive, and it is not clear that crowd sourcing can quickly reach the level of accuracy recommended by the Library of Congress (White 2011: 322). In many respects this discussion is futile because the secrecy of Google's work practices means that not only do we not know how it applies quality assurance procedures, but more importantly, we have no idea whether Google is even working towards Library of Congress levels of accuracy.

This concern extends to metadata generation too. Metadata, which is literally 'data about data', is automatically generated when digital files are created. Of most importance to scholars is the writing of bibliographic metadata, of the type that librarians and archivists have appended to records for centuries. This can only be added manually, which is a daunting task when one considers the vast number of individual and series of digital files that are produced every year. Early studies of the Google Books project suggest that its metadata is not as accurate as similar projects in the public sector and contains more omissions (JISC 2007: 3; Townsend 2007). Even publicly funded projects have been lax in creating bibliographic metadata and generally it is difficult to agree on or enforce common standards (White 2011). All this highlights the naivety of D. Weinberger's (2007: 17 23) earlier comment that the mass generation of metadata can replace classification. But this is not a problem that relates only to classification. Every inaccurate or omitted piece of data in a database limits the amount of information that it can provide. The most disturbing element of this is that while errors in traditional libraries and archives can be easily identified, it is much more difficult to identify errors in hidden databases, especially when they are run by corporations who might see no commercial gain in admitting error.

The long-term sustainability of digital texts cannot be guaranteed and is another reason why we should temper the hubris of the most zealous techno-utopians. There is as yet no answer either to the problem of rapid obsolescence and replacement of digital storage formats and platforms

(Doueihi 2011: 119–121) or to the danger posed by changes in digital storage policies as a result of alterations in the financial or organizational structures of corporations (Vaidhyanathan 2011: 165, 202). One of the reasons why micro-filming was pursued so zealously by librarians in the 1960s and 1970s in the USA was that newspapers were deemed to be acidic and hence at risk of destruction. The result of that campaign was not only the destruction of huge runs of newspapers but also the loss of the information within them, as large amounts of micro-film are now unusable. It would be perverse if the pursuit of universality was used an excuse to destroy original books and other documents, leaving their content solely in digital files whose deterioration or loss would not only imperil the universal digital library but our great existing scholarly paper heritage too.

5. Private *search engines have superseded* public *institutions as the main source of scholarly information*

The point in the previous paragraph about the potential dangers involved in the holding of scholarly information by corporations that could change drastically or even fold within a short period of time will continue to be a salient one in this period of financial instability. There is nothing inherently wrong with private corporations holding this type of information, as the important intellectual role of long-established academic publishers, newspapers, private libraries, archives and educational institutions testifies. But in most countries, the state has acted as a guarantor of educational information through licensing, legal deposit and the public funding of knowledge and educational systems. Furthermore, what makes search engines different from these other private institutions is that their revenue is advertising-based and therefore not directly derived from the information that they disseminate. This makes them less sensitive to the quality of the information that their search engines uncover, unless of course it threatens their advertising revenue. Academic publishers, newspapers, private libraries, archives and educational institutions cannot be so blasé about the quality of their content and hence have more in common with the great public institutions than they do with commercial search engines. For these reasons, commercial search engines should not be the main custodians of scholarly information (see also Vaidhyanathan 2011: 202). Furthermore, despite the national bias of most search engines, they are not as concerned about developing the public sphere as traditional knowledge institutions, the political implications of which will be discussed in Chapter 3.

Conclusion

Let us remind ourselves of the earlier general formula for information-gathering in the modern, democratic nation-state and compare it to that in the age of digital media:

Modern, democratic nation-state

1. Order/classification – cataloguing.
2. Provenance – collected original documents.
3. Access – public libraries, museums and archives widened access.
4. Universal – tried to collect everything.
5. Public institutions hold and control most information.
6. Nation-building – information-gathering institutions linked to power and the nation-state.

Digital media

1. Order/classification – *seen as unimportant.*
2. Provenance – *hard to establish.*
3. Access – universal in theory.
4. Universal – attempt to collect everything.
5. *Increasingly, it is private institutions which hold and control most information.*
6. Nation-building – *not important.*

A number of studies of hypertext reading have shown that students find the lack of narrative structure disorienting (White 2007). There is no reason to believe that this does not continue to be the case. This would suggest that digital media's discouragement of formal classification and imperfect methods of determining provenance has a similar disorienting effect on the user. As was outlined earlier, there is a hidden form of classification online, but it is both intellectually problematic and perverse in that, by privileging what is already popular, it can often expose users to an ever narrower range of information than more academic forms of classification. The attempt to establish some sort of provenance through reputation threatens to degenerate into digital Maoism or mob rule. The lesson, then, is that the Internet needs to learn from academia, the library and the archive. One of the ways in which Google has done this is through the creation of Google Scholar, a search engine for identifying academic papers and their citations. Other initiatives, such as academia.edu, a 'Facebook for academics', which is more structured than Google Scholar, are also promising developments. But it is easy to bash the Internet and the powerful global search engines and neglect the role of the user. The onus is on us too to be less dependent on the search engine for sourcing academic work, which should involve consulting experts, particularly librarians, more often (Halavais 2009: 113).

Focusing on the user leads to a broader point about media literacy. While scholars have been concerned with this for many years (Kubey 1997; Livingstone 2004), the growth of powerful search engines and the advent of the Google Books project in the past decade makes media literacy all the more important. This would begin to address the general lack of intellectual curiosity about the structure and biases of search engines, as illustrated by Deborah Fallows's research in 2005, which found that 68 per cent of users in her sample thought that the search engine was unbiased, while 62 per cent were unaware of the distinction between paid results and those where no money had changed hands (Fallows 2005: i–ii; cited in Van Dijck 2010: 581–582). While one would hope that Internet users have become more savvy since then, initiatives need to be developed to help people make a clearer distinction between authority and popularity (Gillmor 2010; Halavais 2009: 110–111). Ideally, this type of media literacy education should not be solely technical, namely, merely teaching students how best to play the system (Van Dijck 2010: 575). If it does not engage with wider epistemologies about the construction of knowledge, there is a danger that it will merely reinforce the view that the search engine is the best vehicle for delivering the most useful information and it is our task merely to improve the efficiency of our searching, rather than consider alternative strategies (Halavais 2009: 94).

We should not, though, lose sight of the potential of the new research methods that digital media has encouraged. While, as argued earlier, crowd sourcing is not a viable alternative to provenance, it does have a role in quickly and efficiently correcting factual errors. The digitization of archives and library collections has enabled us to find information almost instantaneously, rather than spending days wading through original documents for one reference (White 2011: 318). (This presupposes that institutions are digitizing to Library of Congress standards of accuracy.) And this searching need not only take place within one database, but across a huge number. This can enable the identification of patterns that individuals or even teams of researchers would not be capable of detecting on their own. While there are legal implications in relation to data-mining, it does have the potential to facilitate inductive research for ground-breaking research in diseases and other socially beneficial areas beyond even the wildest dreams of Francis Bacon and the evidential scholarship of the nineteenth century (Vaidhyanathan 2011: 178–179; Van Dijck 2010: 585). The research potential of the search engine makes it all the more important for it to be more transparent about its structure and methodologies (Gillmor 2010; Halavais 2009: 115). While there are many reasons for this opaqueness, primarily commercial confidentiality and their status as private not public organizations, search engines will struggle to retain their credibility as serious facilitators of knowledge construction if they do not make some effort to employ typologies of information-gathering that have elements roughly similar to that of traditional libraries and archives.

While this chapter has focused on knowledge construction, many of the issues raised here have wider political ramifications. One of those issues: the role of social media in altering how we think about individual identity, will be explored in the next chapter. The relationship between knowledge and identity in the age of digital media is profoundly political, the implications of which will be theorized in greater depth in Chapter 3, which is the concluding section of Part I.

2
From the Private to the Public: Online Identity

Introduction

In the previous chapter, it was argued that digital media has, to a certain extent, undermined or, to put it more charitably, refashioned traditional methods of scholarly research. This not only has implications for education but also for politics, for the way in which we construct and maintain our political positions is largely dependent on the quality and range of information that we receive. This is particularly the case where information which was previously monopolized by public, or civic-minded, institutions is now increasingly sourced from commercial search engines and the increasingly privately run Internet, the political implications of which will be explored further in Chapter 3.

As alluded to in the previous chapter, practices like provenance and, in more recent times, crowd sourcing and the highlighting of the importance of reputation or trust, show that identity has always been a significant aspect of information-gathering. The measures that users adopt to establish the authenticity or provenance of an information source cannot neglect the identity of the individual or entity producing it. Theorists of late modernity like Anthony Giddens (1991) and Zygmunt Bauman (2000) have argued that identity is more fluid and not as anchored as it was previously in our respective nation, gender, and class or social roles. Early and mid-1990s Internet theorists like Shelly Turkle (1995) echoed this in their championing of the concept of 'voluntaristic' identities; the idea that one could choose whatever identity one wanted in the online environment. This type of identity is dependent on anonymity and, as this chapter will demonstrate, can be a useful means of exploring identities that we do not or cannot have in the real world. In extreme cases, as the examples of Autumn Edows and Stacey Snyder will demonstrate later, this can have profoundly positive and negative impacts on our offline identities. However, the need to establish reputation and credibility, not least for online shopping and other financial transactions in cyberspace, means that increasingly we are displaying stable online identities which are explicitly linked to our core identities.

The decreasing importance of anonymity online is given concrete expression in the rise of Web 2.0 technologies,[4] most notably social networking. In this more visible online environment, our identities are grounded in our friends in the network. This leads to the very antithesis of anonymity, as many people become exhibitionist in an attempt to stand out from the rest of the crowd. Thus, more of what traditionally we would have regarded as private information about ourselves is voluntarily pushed into the public domain, the implications of which is what this chapter is all about.

Traditional conceptions of identity

Traditionally, identity was conceptualized in relation to one's nationality, gender, religion and/or role within society. Identity also had, and still does to a large extent, identifiably separate private and public elements. While the traditional cultural resources mentioned above cannot be easily screened off from other people, others were generally not privy to our most intimate thoughts and secrets unless we decided to divulge them, which is something we tended only to do to our closest friends and family. But highly industrialized societies of intricate complexity are a challenge to existing conceptions of identity. In what Giddens (1991) and others refer to as the period of 'late modernity', the increasing insecurity and instability of our lives, brought about by rapid economic change, has distanced people from traditional institutions like the church and the state, while also making the 'job for life' rare (Bauman 2000; Beck 1992). Even our attachment to our gender is not as secure as it once was, as experiments with our sexuality and the embracing of a range of different lifestyles have challenged sexual stereotypes. The declining influence of traditional cultural resources was viewed by sociologists in the 1950s and 1960s as problematic, as they believed that it would make people more susceptible to pernicious ideologies with which to fill the vacuum in their lives and/or to lead them to seek order through authoritarianism (Cavanagh 2007: 15).

Others, however, see this as potentially liberating: including early Internet theorists like Shelly Turkle (1995) and Stone (1995). Giddens

(1991) argues that this can enable people to be more 'reflexive' in the way in which they construct and maintain their identities over time (an argument also advanced by Ulrich Beck (1992)). In other words, everyone almost perpetually re-assesses his/her identity (or identities) in light of changing external factors. We could think positively about this in relation to nationalism. A modernist attachment to one's nation might be so secure that it would not be loosened even by the revelation of egregious behaviour by that nation. Conversely, the 'reflexiveness' of late-modernist identity makes it more sensitive and hence, in theory at least, would mean that the individual would likely re-think his/her attachment to his/her nation in the event of this type of behaviour. The work of Erving Goffman illustrates another important aspect of this type of identity, namely self-presentation. Goffman argues that we present ourselves in different ways in varying situations. There is a debate about whether or not these self-presentations constitute separate identities (Cavanagh 2007: 128), but whatever they are, the most important aspect is that individuals must be skilled at managing many different projections of themselves (Palfrey and Gasser 2008: 304). For Giddens (1991), therefore, reflexivity must be augmented by narrative – to remain credible to others, individuals must incorporate all these different self-presentations into an over-arching narrative of the self.

1990s Experimentalism: Voluntarist and post-structuralist online identities

While these trends are significant in changing our conception of public identity (how others view us in relation to our nationality, gender or social role), they do not really call into question the distinction between public and private senses of the self. Neither do they threaten in any significant way our sense of privacy. Theories of online identity that were developed in the World Wide Web's infancy in the early to mid-1990s were constructed in a similar way. These were based largely on experimentation and anonymity. Experimentation is an extreme manifestation of identity's un-anchoring from traditional cultural resources. But on the face of it, as long as these experimental identities are reflexive and contain robust narratives, then why not experiment? In many circumstances, anonymity is an important component of experimental identities. While it is possible to experiment with your identity in full public glare, it is often easier to facilitate this if people do not know much about you. In modern societies with sophisticated forms of record-keeping and surveillance, anonymity is very difficult to achieve. For that reason, the capacity of the networks of the burgeoning World Wide Web in the early 1990s to facilitate communication based on anonymity made them extremely attractive to those who wanted to experiment with their identities.

These experimental identities took two main forms: voluntaristic and post-structuralist. The most prominent of the early proponents of online voluntaristic forms of identity was Shelly Turkle, who argued that 'you are who you pretend to be' (1995: 192; cited in Cavanagh 2007: 120). However, these 'pretend' identities were not developed in isolation but through interaction with others in the network (Lister *et al.* 2003: 348). Thus, not only was this identity multiple but it was also global in its orientation (Cavanagh 2007: 120–121). This form of identity was similar to late-modernist ones, in that it was both reflexive and, despite being fictional, needed to have a coherent narrative in order to be convincing to others. Voluntaristic identity was viewed as liberating, as the anonymity afforded by the Internet enabled people not only to experiment with their identities in a safe environment but also to communicate more effectively as a result of the dropping of the baggage of traditional cultural resources. An example of how this might work is in political discussions, where knowledge of people's nationality or gender is sometimes used against them: if people no longer know that essential information about you then they have no choice but to focus on the strength of your arguments. How this worked in practice will be explored below in the discussion on gender in an online environment.

Mark Poster concurs with many of Turkle's insights about online identity without embracing her voluntarism. His is a more post-structuralist position which questions the whole notion of a rational, unified subjectivity. This new fluid subjectivity is in one respect similar to Giddens's conception of late-modernist identity: 'the salient characteristics of Internet community is the diminution of prevailing hierarchies of race, class and especially gender' (Poster 2001/1997: 268). Poster's reference to the 'Internet community' alludes to the importance of communication to his formulation of modern identity, a classical post-structuralist position. Thus, 'Internet discourse constitutes the subject as the subject fashions him or herself' (Poster 2001/1997: 267). While Poster's post-structuralist position is subtly different from Turkle's (it is neither reflexive nor based on a coherent narrative), both are largely dependent on anonymity, or at the very least people communicating with others without revealing important details about themselves:

In bulletin boards like The Well, people connect with strangers without much of the social baggage that divides and alienates. Without visual cues about gender, age, ethnicity, and social status, conversations open up in directions that otherwise might be avoided. Participants in these virtual communities often express themselves with little inhibition and dialogues flourish and develop quickly.

(Poster 1995a: 90; cited in Lister et al. 2003: 247)

As Poster hinted at above, experimenting with one's gender was one of the most popular forms of identity experimentation in 1990s cyberspace. The case studies below will show how post-structuralist and voluntarist identities worked in practice, providing an insight into the intellectual problems with these concepts.

1990s Internet identities: A case study in gender

Debates in feminism from around the 1970s until the 1990s centred on the opposition between essentialism and anti-essentialism (Stone 2004). In its questioning of essentialism – the idea that women share a common set of natural characteristics that differentiate them from men – feminism not only provided a sophisticated critique of traditional forms of identity construction, but also a seemingly viable alternative. Anti-essentialists considered the male/female binary itself as being at the heart of women's problems. The deconstruction of that binary – like the deconstruction of all other binaries – would thus liberate women. In its emphasis on deconstruction, anti-essentialism echoes much of Poster's post-structuralism. But in its emphasis on political activism, feminism could also be said to be 'voluntaristic' in its promotion of new identities.

Valerie Solana's *SCUM Manifesto*, launched in 1968, drew an explicit association between women's material conditions and technology, seeing hope in the latest innovations:

Life in this society being, at best, an utter bore and no aspect of society being at all relevant to women, there remains to civic-minded, responsible, thrill-seeking females only to overthrow the government, eliminate the money system, institute complete automation and destroy the male sex.

It is now technically possible to reproduce without the aid of males (or, for that matter, females) and to produce only females. We must begin immediately to do so.

(Solanas 2004/1968: 1; cited in Wyatt 2008: 116)

But as the optimism of the 1960s had largely dissipated by the 1980s, other feminist writers took a more jaundiced view of these technologies, believing that men's technical mastery and ownership of them sustained women's marginalization (Wyatt 2008: 116). These essentialist accounts of the effects of technology on gender relations were followed by a more nuanced approach which argued that, regardless of whether or not it was presently 'biased' against women, technology could not be cast aside and thus must be embraced (Haraway 1985). How might this be done? Donna Haraway's answer is that it can be done through the figure of the cyborg. Haraway uses this term to characterize the myriad ways that technology works through us, such that we cannot disentangle it from our core identity. (Here, one can think of pace-makers and other mechanical devices for keeping us alive, or perhaps our seeming dependence on communication technologies like networked computers and smartphones). Accepting that we are cyborgs enables us to use technology in a positive and very political way to bring about changes in society. Perhaps the most important aspect of Haraway's theory of the cyborg in relation to this chapter is the way in which it brings together two seemingly irreconcilable positions: the agency (or 'voluntarism' of Turkle) of feminist theorists of that time like Judith Butler (1990); and the deconstructionism or post-structuralism of anti-essentialist feminists (and Mark Poster) (Wyatt 2008: 118). In this sense, while Haraway was not primarily discussing the Internet (then in existence but years before it was eclipsed by the World Wide Web in the early 1990s), her cyborg prefigured the 1990s debates about gender identity and the Internet.

This raises the question of whether Poster and Turkle are attempting to corral the Internet into the service merely of supporting their existing political ideas. This was not an unusual phenomena at this time, with the argument by writers like George Landow (1992) that the World Wide Web validated post structuralism being well documented (White 2007). This association of particular technologies with specific political ideas was no more aptly illustrated than in the work of Sadie Plant, who argued that the characteristics of the World Wide Web – its networked, not hierarchical, structure and its relative autonomy from direct control – were 'gendered' in favour of women:

[M]any feminists are now finding a wealth of new opportunities, spaces and lines of thought amidst the new complexities of the 'telecoms revolution'. The Internet promises women a network of lines on which to chatter, natter, work and play; virtuality brings a fluidity to identities which once had to be fixed; and multi-media provides a new tactile environment in which women artists can find their space... Complex systems and virtual worlds are not only important because they open spaces for existing women within an already existing culture, but also because of the extent to which they undermine both the world-view and the material reality of two thousand years of patriarchal control.

> (Plant 1996: 170; cited in Wyatt 2008: 119)

But these forms of identity cannot elide the empirical reality of women's continuing marginalization and under-representation:

> Even in Norway, which tops the United Nations' gender-related development index, women's pay is 75 per cent of men's (in many other rich countries, such as Britain, the US and the Netherlands it is about 65 per cent) and the percentage of women who are economically active is only 63 per cent in Norway and still less than 60 per cent in many OECD countries (UNDP 2006). It is largely women who staff the evergrowing 24-hour 'call centres', where shifts are designed to facilitate women's dual role as housewives and workers, and the tasks are repetitive and closely monitored via machines.

> (Wyatt 2008: 120)

Thus, anti-essentialism began to lose its intellectual traction in the 1990s as the solidarity that was needed for successful political campaigns was difficult to achieve with a theory that foreswore the communality of women (Stone 2004: 1). Coincidentally, voluntarist and post-structuralist ideas about online identity also faced heavy criticism around the same time, especially in relation to gender. The main criticism centred on the notion of a disembodied online persona that was completely disassociated from an individual's offline identity:

> The research I have reviewed and the model I have proposed suggest that online groups are often woven into the fabric of off-line life rather than set in opposition to it. The evidence includes the pervasiveness of off-line contexts in online interaction and the movement of online relationships off line.

> (Baym 1998: 63; cited in Lister *et al.* 2003: 168)

These theoretical challenges to online voluntarist or post-structuralist identities were accompanied by outrage over some examples of unethical experimentation with online identity. The first example was that of Julie Graham, a self-described disabled and disfigured psychiatrist who practised online (Stone 1995; cited in Lister *et al.* 2003: 168–169). The interactions between Julie and her 'clients' took place within a feminist paradigm and provided great succour to a number of women over a couple of years. However, 'Julie' was actually a male therapist by the name of Sanford Lewin. When this was revealed, it caused much distress to those who believed that they had been deceived (Lister *et al.* 2003: 169). The second example involved an online rapist named Mr Bungle. In the online community, LamdaMoo, in which he operated, Mr Bungle's activities caused widespread disgust (Dibbell 2001/1993). Two things stand out in this example: the sense of outrage suggested that this was not merely a online issue but a real world one too; in the discussion that followed, participants suggested solutions that were similar to the type of political approaches that one would take to address sexual violence in the real world.

What these two examples demonstrate is that while it is technically feasible to create disembodied and other experimental identities online, participants tend for the most part to act as if these identities are 'real'. While it is possible for participants to accept up-to-a-point new identities in environments, like gaming and role-playing, this is much less palatable in discussion fora. (The case of Mr Bungle shows that even in role-playing and gaming environments people are often expected to play by the rules and not be morally dissolute.) The reason this is so is that, contrary to voluntarist or post-structuralist accounts, people's online identities usually cannot be disentangled from their real personas. This demonstrates the limits of the concept of late-modernist identity, in the sense that reflexivity and a sense of narrative can only augment, not completely replace, traditional cultural resources.

Hyper-identity? Online and offline identity in the twenty-first century

Allison Cavanagh (2007: 121–125) has attempted to theorize this complexity in terms of 'hyper-identity'. The characterization of the Internet as a network is central to her thesis, a network that can only function properly with the existence of stable identities. Cavanagh has two types of hyper-identity in mind here: one – reputational – which is commercial in orientation; the other – productive – which enables individuals to attract attention to themselves. Both forms of identity are based on trust, something sadly lacking in the two examples of online deception shown above.

What Cavanagh refers to as a 'productive' (2007: 122) form of identity can be linked to authority or provenance. That is to say, users gauge the degree to which online information is authentic with reference to the credentials of the creator. If, therefore, someone writing a blog or creating other types of online content wants to be taken seriously, he/she must find a means of convincing the potential audience of his/her credibility. Impersonation (potential or real) and anonymity are not inherently trustworthy characteristics. So, in order to create a credible persona, a person must not only be not anonymous, but also must convince others that he/she is not impersonating someone else. One of the most effective ways of doing this is by placing oneself within a network of relationships. This is easy to do offline, but not so easy online. That is why social media is so important to the construction of this type of online identity, in that it facilitates these links. At the same time, this network of relationships must be grounded in real world identities for it to have a credible foundation (Caldas-Coulthard 2005). In this sense, contemporary online identity has moved a long way from the voluntarism or post-structuralism of its 1990s forebears.

While it is feasible to apply Giddens's (1991) ideas about self-reflexive identities to online identity, there is one crucial extension to his thesis that must be taken into consideration. Giddens's late-modernist identity is forged within specific contexts: while we might have more than one narrative, each is related to a (different) particular context (we might be bankers during the week but punks at the weekend). But our online identities are constructed before a potentially unlimited audience (Papacharissi 2010: 142). In the competition for a slice of this audience, we are likely to become more extrovert and more willing to divulge more information about ourselves than we are willing to do in other environments (Cavanagh 2007: 124); Caldas-Coulthard (2005) has observed this effect, particularly the blurring between previously private and public information, in the construction of academics' websites.

This means that, as well as our online identities serving as vehicles for fantasy and experimentation, the difficulty of disentangling these identities from our real life personas means that increasingly what we do online is shaping our offline selves. An example of this is the case of the schoolgirl Jessica Hunter, whose teenage shyness and isolation led her to creating an exotic online persona 'Autumn Edows' (Palfrey and Gasser 2008: 306–307; PBS Frontline 2010).[5] Her gothic alter-ego became so popular online that this gave Hunter much more confidence in herself, transforming her whole life (PBS Frontline 2010). The more negative story of a 25-year-old trainee teacher called Stacy Snyder is related by Mayer-Schönberger (2009). Despite passing all her teacher training exams, Snyder's university refused to give her a certificate after viewing a photograph on MySpace of her drinking alcohol while wearing a pirate's hat (Mayer-Schönberger 2009: 1–2). This highlights the danger of posting what are images of essentially private behaviour on public platforms. As in Hunter's example above, Snyder's online persona – a self-proclaimed 'drunken pirate' (Mayer-Schönberger 2009: 1) – impacted significantly, in this case malignly, on her real life.

Cavanagh's (2007: 121–122) other form of hyper-identity, one which pivots on commerce, says much about cyberspace's mutation from an early to mid-1990s public-oriented Internet to the commercially dominated World Wide Web that we know today. When the first major e-commerce companies like eBay were established, their philosophy embodied the collaborative culture of the early Internet in their belief that sellers would not attempt to cheat buyers. Such was the prevalence of unethical behaviour in eBay's fraught first few weeks that it swiftly disabused the company of its sanguine attitude to its users. Thus the 'reputation system' was born (Shirky 2010: 177). This system works on the basic premise that vendors are rated by those who have bought goods from them. A vendor who has collected many positive reviews over a number of years will therefore be trusted more than someone who has not received many reviews or has many negative ones. The system, therefore, penalizes the con man and rewards trustworthy sellers – by up to 8 per cent in terms of selling price, according to Resnick *et al.* (2006). This has become common practice in online selling, perhaps best exemplified by Amazon.

This, then, is a very public form of identity, a significant departure from what was promised by the voluntarism of 1990s Internet scholars. While it would be impossible, though, for any identity to exist without reference to its external environment, modernist and late-modernist identities presuppose that we have a certain degree of personal autonomy. This personal autonomy was buttressed by the concept of privacy, the idea that there were many aspects of people's lives that they did not want to divulge to others beyond their closest friends and family. The development of the interior, private self has been attributed to the invention of the printing press in fifteenth-century Europe and the subsequent rise in literacy (Ong 1982). Walter Ong (1982) and Marshall McLuhan (1964) argued that what they termed 'electric media' would so transform literate societies that they would share many of the characteristics of the oral societies that preceded them. Neither theorist was discussing digital media, but they are considered prophetic in the context of the capacity of social media to challenge our long-held concept of privacy to such an extent that it does not seem entirely improbable to argue that it now has a similarly ancillary status to that which it had in fully oral societies. The experience of Stacy Snyder would seem to suggest that, despite the impact of globalization, the way in which our private behaviour is surveilled online is similar to that which we would expect in a small village; to use a McLuhanite phrase, we now live in a 'global village'. (A fuller discussion of McLuhan and Ong's work on literacy will take place in Chapter 6's discussion on reading digital media.)

While not citing them directly, Ong and McLuhan's idea that digital media technologies will make humans more social and attenuate individuality and privacy is celebrated by many contemporary Internet theorists and practitioners (see Keen's (2012) reporting of the views of technology writer Jeff Jarvis and social media guru Reid Hoffman). In a speech in 2011, Jeff Jarvis asserted in McLuhanite terms that social media is returning us to an oral society (Keen 2012: 52). Furthermore, our modern concept of privacy is being replaced by 'publicness', the idea that we share virtually everything. Jarvis believes that this is to be welcomed as it will make us more tolerant – there will no longer be any dirty secrets or taboos (Keen 2012: 52). And Jarvis cannot be accused of not practicing what he preaches, as he informed the world of his prostate cancer in a blog in April 2009. (This practice is quite common among social media champions, an example of which was Guy Kewney's chronicling of his impending death in 2010 on social media) (Keen 2012: 26). Not only do these writers and business leaders practice what they preach, they also have difficulty understanding why others would not want to do the same. This is illustrated by Reid Hoffman's incredulity when Andrew Keen (2012: 7) said to him: 'Let's face it, Reid, some people just don't want to be connected.'

In relation to the concept of hyper-identity, what works for already rich and famous writers and business leaders like those above might not work for the average citizen, as the case of Stacy Snyder demonstrates. Nonetheless, as Lanier (2011: 4) argues, these new technologies are involved in a form of social engineering, even if it works more subtly than Jarvis and Hoffman's suggestions. One of the ways in which it arguably has done this is in the shaping of the 'networked self'. Papacharissi (2010: 138–144) has developed the concept of a networked self that enables individuals to access public information from private platforms. This means that individuals are able, if they so wish, to preserve privacy and exercise some autonomy as they travel through cyberspace. This may involve the construction of many different identities (Palfrey and Gasser 2008: 22) or, more likely, a multiplicity of different elements making up the one, unitary identity. In this sense, this sounds similar to late-modernist identity. Unlike late-modernist conceptions of identity, however, it is difficult to keep your different 'faces' separate and so the unitary aspect of one's identity (the need to have a coherent, over-arching narrative which spans both our online and offline existence) is the most significant factor. In other words, the development of the networked self implies some sort of voluntarism, but this is restricted by the singularity of identity – online identity is inextricably linked to our real life selves. If identities, then, are being increasingly crafted in an online environment of a potentially worldwide audience, this will impact on construction of the offline self as well.

In what ways, then, is the networked self fashioned? A cursory reading of Papacharissi (2010: 138–144) would suggest that this is done in a skilful, carefully constructed manner, divulging to complete strangers only that which the individual would wish to be put in the public domain, while revealing more intimate thoughts only on social media platforms like Facebook and MySpace which restrict information flow to one's 'friends'. However, as Papacharissi (2010: 142–143) herself acknowledges, managing one's identity is a little more complex than that, as we are constantly being asked to make decisions in relation to it as we travel through cyberspace – just one wrong micro-decision could result in us divulging more information than we intended. This is made particularly difficult by the default settings of many social networking platforms, which are biased towards information-sharing rather than privacy, the pressure on people in social networks to divulge information both in order to make friends and to stand out from the crowd, and the fact that some people are simply not as skilled at navigating online as others (Palfrey and Gasser 2008: 17–37; Papacharissi 2010: 143). Added to this, as Stacy Snyder found to her cost, is that we are not as concerned about our privacy when we are younger as we are later in life.

There is another reason why the notion of the networked identity is unattractive to some. Our private selves enable us to retain a mystique which is at the very heart of our identities:

> I am dreaming of a Web that caters to a person who no longer exists. A private person, a person who is a mystery, to the world and – which is more important – to herself. Person as mystery: This idea of personhood is certainly changing, perhaps has already changed.
>
> (Smith 2010; cited in Keen 2012: 184–185)

In the same vein, Lanier (2011: 48) says of online identity:

> The strangeness is being leached away by the mush-making process. Individual web pages as they first appeared in the early 1990s had the flavour of personhood. MySpace preserved some of that flavour, though a process of regularized formatting had begun. Facebook went further, organizing people into multiple-choice identities, while Wikipedia seeks to erase point of view entirely.

This line of argument is reinforced when we think of with whom we are sharing our most intimate thoughts online. Offline, we would restrict this information to our closest friends and family, but what about online? A 2011 Pew Center study found that the average Facebook user has 229 'friends' (Hampton *et al.* 2011; cited in Keen 2012: 173). However, Oxford University Professor Robin Dunbar – famous for 'Dunbar's Number' which calculates our optimum number of friends – believes that human beings are able to keep a detailed track of the lives of only around 150 individuals (Keen 2012: 174). Clearly, then, many Facebook 'friends' are not friends in the conventional sense, but these are the people with whom we are sharing some of the most intimate aspects of our lives. Therefore, even many of those who try to construct their networked selves by making a carefully calibrated distinction between their public and private pronouncements are making the latter to some people who are not proper friends, some of whom they have never met offline. While the idea of 'publicness' as a means of constructing identity has obvious problems, there is no doubt that our private spaces online are becoming attenuated and/or infiltrated by interlopers masquerading as 'friends', thus blurring the boundary between our public and private selves.

Tribes: More traditional forms of identity online

Even if it is not a fully voluntary process, the construction of identity in the modern world appears to a certain extent to enable people to move away from the restrictions imposed by traditional cultural resources, like nationality, gender and class. But, as the examples of psychiatrist 'Julie Graham' and Mr Bungle showed, this rather upbeat view of the diminishing influence of traditional cultural resources is contestable. As Wynn and Katz (1997: 299–300) point out, the idea that this trend (if it indeed exists) is liberating is at odds with Durkheim's empirical research on suicide, which he linked to the anomie of modern life. In Durkheim's view, these traditional cultural anchors provide structure to some people and enable them to better deal with the complexities of modern life, and hence their removal brings not liberation but dislocation and disorientation. Maffesoli (1996) similarly argues that mass society has undermined what he refers to as the 'solidarity' that traditional cultural resources gave individuals. In order to re-create this solidarity, Maffesoli argues that people organize themselves into 'tribes'. But while these tribes represent attempts to re-create solidarity, they are usually not intended to re-create traditional cultural resources. Rather, they can usually be thought of as entirely new formations based on personal tastes and interests. Obvious examples of tribes include adherence to the latest fashions in clothing, and the link between the tribe and consumerism will be explored further later. But a tribe can consist of people whose commonality is non-commercial, perhaps to a political cause or personal hobby.

Maffesoli's book was written in his own language (French) in 1988, long before the Internet morphed into the World Wide Web. Despite that, it is easy to identify similarities between his concepts and some more contemporary writers. Cass Sunstein (2007) believes that the Internet accentuates this trend by making it easier for people to find members of their own particular tribe. While forming tribes based on particular interests was not insurmountable in the days before the World Wide Web really took off, the opportunity that that platform affords for bringing together fellow believers of even the most esoteric cause is unprecedented. People who in an earlier time might have found it difficult to meet others with the same esoteric views as them, now can use the Internet to facilitate daily discussions with like-minded people in many different parts of the world (Sunstein 2007: 53–54). A related and similar phenomenon is the rise of the so-called 'Daily Me', or personalized news (Sunstein 2007: 1–18). This is news that is filtered to suit the particular interests of one viewer or reader – you! Thus, rather than listening to news programmes in their entirety as we might have done 20 years ago, it is now possible to receive only those stories and viewpoints that reflect our own interests and prejudices, with those that expose us to a more eclectic range of interests and which might challenge our own views being effectively filtered out. If we spend a disproportionate amount of our time conversing with like-minded people and being exposed to a range of information which is not only ever narrower but largely reflects our own prejudices, there is a danger that our views will become more polarized.

Sunstein's perception of information-gathering online is similar to the ideas expressed in Chapter 1 about the tendency of search engines to narrow our range of information sources. But what makes Sunstein's (2007: 67–69) argument a potentially more potent one is the role of identity in the construction of knowledge. This is because our views are shaped by our perception of ourselves, our identities. In other words, if like-minded people agree with us on a particular issue, then our opinion will likely be strengthened; if those same people disagree with us, then we might well re-evaluate our position. Similarly, if you find yourself agreeing with 'people not like you', then you might conclude that you are wrong to hold a particular opinion (Sunstein 2007: 69). Sunstein argues that this echo-chamber – like-minded people reinforcing each other's views – can make people more extreme and thus lead to polarization. Sunstein's idea reinforces the argument that even though the information-gathering regime to which we are subject nowadays is more malleable in structure, filtering still takes place. And the method of filtering is very much dependent on our sense of identity. While many of these tribes take new forms, as Sunstein has demonstrated, many of them are based on the traditional cultural resources that were supposedly weakened in late modernity. Along with Vaidhyanathan's (2011: 138–139) observation in the previous chapter that Google searches are becoming increasing location-specific, this suggests that for all the discussion about the opportunities afforded by the Internet for creating new forms of identity, more traditional ones are flourishing in this environment and continuing to shape the way in which we see the world around us.

As well as being useful to those who want to find people who share their interests, tribes are also a boon for advertisers. In the mass media environment, companies sold a narrow range of content to the mass audience. The Internet has given advertisers the capacity to more precisely target their messages to segments of the market. A crucial part of any advertising strategy will therefore involve finding those segments and the best message for each of them. This is the crux of Seth Godin's (2008) book on tribes. While Godin talks about tribes in a range of situations, both commercial and non-commercial, his emphasis is on their impact on modern marketing: 'Today, marketing is about engaging with the tribe and delivering products and services with stories that spread' (Godin 2008: 13). While those in marketing have always targeted tribes – here, one thinks of the commercialization of punk and the association of certain clothing brands with football hooligans – nowadays there are many more tribes and the technology, through innovations like cookies and online transactions, to more quickly and accurately identify members. It is easy to think of examples of this, Amazon's emailing of people to suggest book purchases based on their recent buying or surfing activities being a particularly clever use of this tactic.

While the last example would seem not to be that intrusive and, in many cases, helpful – we need much more information when deciding which book to buy rather than when we are buying something whose qualities are more transparent – there is a troubling aspect of the exploitation of these new forms of identity for marketing purposes. Andrew Keen (2012) argues that those very social networks that enable us to connect with other people better than ever before and hence, to use Maffesoli and Godin's terminology, to better facilitate the formation of tribes, are also exploiting their members. How do they do this? Through storing huge amounts of data on each member, a veritable goldmine of information which is then sold to advertisers. Members think that they are getting something for free, but what they are essentially doing is paying for their free ride by acquiescing in their own surveillance (Lanier 2013). This might be a Faustian bargain which the vast majority of Internet users are willing to accept, but we should not pretend, then, that the new identities forged in this environment are any less restricted than identities based on traditional cultural resources.

Conclusion

During this chapter, we have seen how early Internet theorists argued that the medium generated new types of identity, be they voluntarist or post-structuralist. These, though, proved to be both epistemologically problematical and impractical in the long run. The opportunity to experiment with their identities was attractive to many people, but the increasing need to ground these in their real lives, both for emotional and transactional purposes, meant that hyper-identity became a better conceptual label to describe this change. What is, though, unique about online identities is the way in which they can be constructed in relation to an unlimited audience rather than the relatively small circle of family and friends through which we fashioned identities in the pre-Internet era. This enables us to emphasize those aspects of our identity that are based on taste and interest more than we could have done in the past – the Internet enables us to find people with similar interests and tastes, however esoteric our proclivities and however far away those fellow tribe members live from us. This might mean that those tastes and interests become reinforced by constant encouragement by like-minded people and through our need to 'perform' online to attract attention. There is a danger that we become more isolated from our immediate offline families, friends and colleagues as choosing our company on the basis of shared interests becomes habitual – this also lessens our exposure to information that is outside our interests and worldview and thus can make us less rounded intellectually and emotionally.

In this sense, we can see that identity is an important factor in the way in which we construct knowledge; changes in identity will, then, impact on knowledge construction. What can be said generally about the move away from identities based on individuality (literate culture) and anonymity (1990s Internet cultures) to ones that are more public in orientation (social networking) is that it subtly alters the relationship between the private and public spheres as traditionally conceived in political theory. We saw in the first chapter how information that has traditionally been monopolized by the public sphere is now increasingly being stored by and mediated through private companies. Conversely, this chapter has demonstrated that the information we have about our own lives (our private information) has become increasingly public (albeit much of it monetized by private companies). Jaron Lanier (2013: 10) describes how this works in social networking: 'cryptography for techies and massive spying on others'. Sunstein argues in relation to online tribes that their tendency to encourage echo-chambers and polarization requires a political response. So too does the shifting and complex relationship between those public and private spheres which are integral to modern constitutional politics. In her book *A Private Sphere*, Papacharissi (2010) argues that, in their framing of their analyses through the concept of the public sphere, Internet theorists are unable adequately to explain the paradigmatic shifts that have occurred in people's engagement in politics through online networks. Papacharissi and Sunstein's views will be interrogated in the next chapter as part of an attempt to theorize whether and to what extent these changes in identity formation and knowledge construction have ushered in a new form of politics.

3
Digital Media and Politics in the Liberal Democratic State

Introduction

The previous two chapters have considered the implications of the growth of digital media for the way in which we seek information to make sense of the world around us and conceive identity. The questions this disquisition raises are profoundly political in orientation. Unless we employ a crude form of technological determinism, we cannot discuss these issues without analysing the relationship between digital media and wider society. A crucial element of this type of analysis involves determining to what extent these changes in the behaviour of the users of this technology are *determined* by the intrinsic properties of that technology, wider social trends, or a combination of the two. It will also involve assessing the extent to which changes in the way in which we interact with digital media are politically or socially significant, or whether they are marginal to those concerns. By way of example, we might consider the concept of online identity. The previous chapter has identified significant changes in the way in which our contemporary identity is constructed as a result of our interaction with digital media. But is digital media ultimately responsible for these changes or are they more likely a result of contemporary social and political trends? And are these changes really that socially and politically significant? What if our increased willingness to experiment with our identities in and of itself has little impact on wider politics or social life?

This chapter will attempt to answer these broad questions by placing the issues of identity and information-gathering within the context of contemporary political theory. Given that it is beyond the scope of this book to discuss all aspects of contemporary political theory, the focus will be on the role of the state in liberal democracies and the public sphere. (An analysis of states, mainly in the developing world, that are not liberal democracies, will be the subject of Chapter 9). As the chapter will go on to demonstrate, there is a widespread perception that the liberal democratic model is under threat, two of whose proximate causes are argued to be increasing globalization and/or the proliferation of digital media. Given that most of us in the western world live within this type of political construct and many more in other parts of the world desire to do so, the fate of the liberal democratic state is of crucial importance to contemporary politics and social life. In relation to this, the chapter will interrogate three broad questions. First, is the liberal democratic model under threat? If so, what role has digital media played in undermining it? Can digital media help us to fashion an alternative viable form of politics or, at the very least, a reconfiguration of existing models? This normative approach will provide a context for some of the later discussions of the actual use of digital media in political and social campaigns in Chapter 7.

The chapter will begin by analysing the characteristics of the modern liberal democratic state and establishing the importance of the role of identity formation and re-formation. The main normative model employed in these discussions is the 'public sphere'. This was chosen because, as the enfolding discussion will show, however intellectually problematic we view the concept of the public sphere, it is difficult to talk about politics and the social in contemporary liberal democratic states without acknowledging the centrality of identifiable public and private spheres, however so defined. This also ties into arguments in Chapter 1 about how we construct knowledge, principally because we have traditionally gathered information about social and political issues from institutions, like the mass media, within the public sphere. That the emergence of global digital networks has, depending on one's conception of the public and the private, seemingly contributed to making those spheres harder to define, delineate or save from perpetual reconfiguration is all the more reason for a study of this kind (see Dahlberg 2007a; Papacharissi 2010).

The liberal democratic state: A crisis of legitimation?

The liberal democratic nation-state as locus of political power and legitimacy is something that those of us who have lived in the so-called developed world have been subject to at some stage in our lives. The increasing globalization of economics and culture over the past few decades seems, though, to pose an existential threat to this model. Many of humankind's biggest problems, like preventing further proliferation of nuclear and biological weapons, combatting international crime and terrorism, reversing global warming and mitigating the excesses of global capitalism, are far too complex to be tackled by individual nation-states. For that reason, there has been an increasing tendency for states to cede sovereignty in some areas of decision-making, to supranational institutions (as many European states have done in the guise of the European Union and most nations have done in global institutions like the World Trade Organization), to be more subject, through their embeddedness in a global economic order, to decisions made by powerful actors elsewhere (witness the role of the G8 and the G20 in re-stabilizing the global economy after the financial crash of 2008) and to be susceptible to environmental problems emanating from other nation-states. The subject of this book, digital media, exacerbates these tendencies through its proliferation of information from and to all parts of the world, thus weakening the media's role in developing and reinforcing national cultures. Added to that is pressure from below in the form of demands for greater regional autonomy if not outright secession in many states, like the UK, Spain and Canada. Citizens of even the most liberal and democratic countries could be forgiven for thinking that the nation-state faces a crisis of legitimation. The fall in participation in national elections in the oldest democracies since the 1970s (Moog, S. and Olle, E. in Castells 2010b: 410) chimes with the claim that: 'The large majority of citizens around the world despise their representatives and do not trust their political institutions' (Castells 2010b: xxxiii).

If, as suggested above, the liberal democratic nation-state is in crisis, how might this problem be addressed? A superficially logical answer would be to further cede sovereignty to supranational institutions. But that presupposes that this option appeals to citizens. Even in probably the most successful and ambitious pooling of sovereignty, namely the European Union, attempts at further integration have been slowed by the rejection of its treaties in national referenda in some member states. Pippa Norris's analysis of the World Values Survey in 1990–1991 and 1995–1997 illustrates that people's self-ascription is based on locality and nationality, rather than a sense of global citizenry. Only 15 per cent of those surveyed described themselves primarily as world citizens, with a further 2 per cent being ascribed this way exclusively. For 38 per cent of people, the nation was the primary source of their territorial identity, while 47 per cent identified most closely with their locality or region (Norris 2000; cited in Castells 2010b: 335). Liberal theorist Will Kymlicka (2002: 269–270) provides a cogent explanation for this:

> Cosmopolitan critics of liberal nationalism often write as if the natural or spontaneous tendency of human beings is to feel sympathy for all other humans around the world, and to be willing to make sacrifices for them. Nationalism, on this view, artificially restricts our natural inclination to conceive of and pursue justice globally. Liberal nationalists, on the other hand, argue that people's natural or spontaneous sympathies are often very narrow, far narrower than the nation-state. Historically, the sort of redistribution that people have willingly accepted has generally been limited to their own kin group and/or co-religionists. This was an insufficient basis for justice in modern states, since citizens often differ in their ethnicity and religion. Nation-building, then, was a way of 'artificially' extending people's restricted sympathies to include all co-citizens, even those with a different ethnic background, religious confession, or way of life. If we were to reject the moral salience of nationhood, as cosmopolitan critics propose, the worry is that this would lead, not to the expansion of our moral sentiments to include foreigners, but rather to the restriction of our moral sentiments back to kin and confession, as was the historic norm.
>
> For liberal nationalists, then, the extension of moral concern to co-nationals was an important and fragile historic achievement, and should not be abandoned on the naïve expectation that people's natural sympathies are global in scope. That is not to say we should reject all concerns for global justice, but rather should move towards global justice by building upon, rather than tearing down, the achievements of liberal nationalism. How the centrality of nationhood can be maintained while simultaneously working towards a more global sense of justice is not easy to see, and much work remains to be done in sorting out the relations between liberal egalitarianism, liberal nationalism, and cosmopolitanism.

Even cosmopolitan critics of the kind that Kymlicka cites above agree that the nation-state remains the most viable mechanism for providing entitlements and redistributing resources from the richest to the poorest members of society (Appiah 2007: 163; Parekh 2008: 54).

But, given the challenges to it identified above, which normative model of the nation-state is the most viable in our increasingly networked and globalized world? The liberal nationalist model, as expounded by Kymlicka above, is based on the promotion of a 'thin' type of national solidarity based on a common language and liberal public institutions. Liberal nationalists argue that not requiring its citizens to adhere to a specific way of life, usually cultural or religious in orientation in non-liberal nation-states, makes it more likely that those citizens will give their allegiance to the state. (In a similar formulation, Parekh (2008: 79) argues that citizens can share broadly defined economic and social goals rather than focus on what constitutes national identity). There are two main arguments against this liberal nationalist conception of the modern state. The first is that over time states will tend to 'thicken' national identity, in other words to make it more culturally or religiously specific in orientation (Kymlicka 2002: 265–266). Whatever liberal nationalists say, the nation-state is and has always been by and large constructed around the values of the dominant ethnic, cultural and/or religious group within each territory (McGarry and O'Leary 1995: 130).

For communitarians – who pose the most credible intellectual challenge to liberal nationalists – citizens' ideas of individual liberty must be grounded in the social (Sandel 1998). (Not surprisingly, Kymlicka (2002: 270–271) challenges the idea that liberals are not concerned with the grounding of rights in the social and thus argues that the concern about liberal 'atomism' is a 'straw man'.) As might seem obvious from their name, and from their emphasis on the social, communitarians' political project involves either a reconstruction of traditional communities or the creation of new ones. This is imperative because citizens can only be bound together by a sense of the 'common good', which is grounded in the wider community (Kymlicka 2002; Sandel 1998).

But communitarianism is not really able to overcome the same sort of contradictions to which liberal nationalism is subject. As suggested above, the 'common good' cannot avoid incorporating some of the 'thick' elements – religion, culture, ethnicity – into it. This is largely because (as will be explored in more detail later in this chapter), identity based on traditional cultural resources has, contrary to some of the theories of late-modernist identity explored in the previous chapter, not lost its lustre; it could even be argued that globalization has made it *more,* rather than *less* popular as a means of self-ascription (Castells 2010b). And in our modern multicultural societies, any incorporation of these elements into a conception of the common good will alienate many citizens. One way of overcoming this is by institutionalizing communalism, in other words, accepting that policies should be formulated primarily for communities rather than for individuals. This is a popular course of action in divided societies, as can be seen in the power-sharing (between the predominantly Catholic Irish nationalists and predominantly Protestant Ulster unionists) constitutional, political and legislative architecture of Northern Ireland. But this type of constitutional arrangement is not appropriate for most liberal democracies. That being so, the logic of communitarianism is that it can only really flourish at the local level (Kymlicka 2002: 266–267).

As the statistics cited by Norris above suggest, it is the local or regional, rather than the supranational, that provides the most potential for a reconstruction of politics in liberal democratic states. Castells (2010b) argues too that the crisis in the legitimacy of the nation-state has resulted in central governments around the world ceding more power to regional and local governments. (This can be a double-edged sword for regional and local governments, as the acquisition of more power also brings with it some of the questions of legitimacy that hitherto were the almost exclusive concern of central governments.) This tendency towards the devolution of power could also take place at the community level, as communitarians would argue. This is consistent with the view of Castells (2010b) and Parekh (2008) that communal identity is as important now as it ever was. For Parekh (2008), a form of what we might term strategic essentialism enables groups to overcome their marginalization through solidarity, an essential feature of which is the writing of their *own* history, rather than that of the universalized narrative based on the story of the dominant culture. An example of this positive form of identity politics is feminism, which has been successful in challenging patriarchy through solidarity campaigns and the creation of an academic canon to rival the so-called universal (male-centred) version of history. Indeed, Castells (2010b: 11–12) argues that the logic of the global political construct he refers to as the network society (of which more later), its undermining of civil society and the nation-state, makes communal resistance (resistance based on identity rather than value-based movements like socialism) the most popular and effective means of opposition to it. But a type of politics based on communal identities imperils the redistribution project for which the nation-state is the most effective mechanism. As Kymlicka (2002: 268) argues:

> We may be able to sustain redistribution *within* local communities or neighbourhoods that share a common conception of the good, but the most urgent injustices require redistribution *across*

communities – e.g. from white suburbanites to black inner-city families, or from wealthy Silicon Valley to poor Appalachia. Local communitarian politics cannot help us here. This requires a source of solidarity which is not local, and not grounded on a shared conception of the good [emphasis in the original].

Added to this is Sen's (2007: 32–36) criticism of the relativism that grounding values in discrete communities can tend towards. In this sentiment, Sen allies himself with Kymlicka's (2002) argument above about the difficulties of redistribution in such political arrangements. This is compounded by Castells's (2010b) observation that global net-worked capitalism has undermined welfare states by driving production and wage costs downwards.

Parekh's (2008: 53–55) formulating of a bifocal form of justice is an attempt to address these concerns. In this model, the confidence that a communal identity can bestow on marginalized groups can propel them into the mainstream, rather than fracture society along particular-istic lines – here one thinks of the mainstreaming of the LBGT (lesbian, bisexual, gay and transgender) community in its successful campaign to institutionalize gay marriage, an essentially conservative goal for a previously radical movement. In his tripartite schema of modern iden-tity, Castells (2010b: 6–12) considers a similar phenomenon in the movement of some *resistant identities* (communal identities of marginal groups) to *project identities* (attempts to change the whole of society based on the values of the resistant identity) to *legitimizing identity* (suc-cessful, in that the previously resistant identity is now part of the official state ideology). We might view feminism as an example of a resistant identity which has transformed liberal democratic societies. However, this is an exceptional case based on an identity whose influence has not caused as much disruption as the triumph of other resistance identities, especially those centred on race or religion, undoubtedly would. Indeed, the acceptance of feminists' and gay activists' norms could be consid-ered commensurate with liberal democratic ideals; it is hard to see how some of the other resistant identities that Castells cites, like for instance Christian or Islamic fundamentalism, could be similarly embraced by liberal society.

Since this is not primarily a book about liberal philosophy, this argu-ment will not be extended beyond this point, save to re-emphasize the continuing difficulty for the liberal democratic state in constructing race/gender/culture-blind policies in an era when communal identity is arguably the primary means of people's self-ascription, and where globalization has moved millions of people into states that they were not born into. Instead, I will continue with a discussion which I believe is more pertinent to my arguments about digital media. While all theorists above have different views about the wisdom of what we might crudely label identity politics, none argue against the proposition that identity plays a critical role in the politics of modern liberal democratic societies. Furthermore, the most influential of these identities are those based on traditional cultural markers, like nationality, gender and religion. This accords with the analysis in Chapter 2 of the failure of post-structuralist and voluntarist online identities, thus suggesting that theorists of late-modern identity like Giddens and Bauman were over-optimistic (or, depending on your perspective, pessimistic) in their championing of identities less anchored in traditional cultural resources.

Digital media and identity formation/re-formation in liberal democratic societies

What role does digital media play in identity formation and re-formation in the liberal democratic state? First, it reinforces some of the trends outlined here. The rise of the type of online tribalism identified in the previous chapter by Cass Sunstein makes the fashioning of communal solidarity easier. An example, albeit an extreme one, of this is the role of the World Wide Web, in its capacity to bring together disaffected Muslims from all corners of the globe, in the rise of Islamic fundamentalism (Castells 2010b). That said, it can also bind together those who are motivated by concerns that are not communal but based on universal political values. Here, one thinks of the global environmentalist movement, the influence of which is proof that identity politics is not the only game in town in our networked societies.

Chapter 2 also highlighted the importance, more generally, of the role of digital media in identity formation in contemporary societies. There, we saw how online identities cannot be completely separated from offline identities. This explains why we cannot simply fashion completely new identities on the hoof: online identities must have some grounding in our real ones. Conversely, our real life identities cannot insulate themselves from what we do online; ergo, the more time we spend online, the more it influences our sense of identity offline. Without wishing to repeat clichés about teenagers' supposed mastery of digital technology, this is particularly important for people who have not experienced a world in which digital media is not prominent and in which they are not immersed (Palfrey and Gasser 2008). If identity is of such importance to the citizens of what continues to be the most viable political entity, the liberal democratic state, then the site (increasingly digital media) of its formation and experimentation is of interest to us all.

It should be acknowledged that though the nation-state continues to be the most viable model within which people practice politics, like identity, it too has been altered by the rise of digital media. While many of the changes to the nation-state could be said to be the result mainly of globalization generally, it would be a mistake to underplay the role of communication in that process. Manuel Castells's theory of the network society, explicated articulately in a number of books and papers, demonstrates the centrality of communication technologies to contemporary globalization. While neither globalization nor its particular association with communication are new phenomena, there has been a qualitative change in the capacity of communication networks to send information at speed, such that they have almost reached global instantaneity. (Castells (2010a) argues that these sophisticated communication networks are central to the largely successful attempt to re-structure the modern state in line with the dictates of global networked capitalism, something that will be explored in the second part of this book.)

We can observe other changes too, which can be related to the latter section of Chapter 2, which suggest that modern online identities are constituted in public and in the face of a potentially unlimited audience. If we accept the premise that online identity construction does not deviate too far from the way in which we think about identity in actual society, then is it true to say that the increasingly public nature of what were previously private identities is the new norm? Related to that is a more general question about the relationship between the public and the private in contemporary liberal democracies. Modern societies, particularly those which are fully democratic, are characterized by the existence of distinctly separate public and private spheres which provide norms that enable citizens to navigate their way through civic life. The public sphere (which theorists of these spheres tend to give most attention to) is closely linked to civil society (or the civic), and both in turn are related to political legitimacy (Downey and Fenton 2003; Habermas 1991/1962; Kymlicka 2002: 261). If Castells's earlier comment that the informational society has undermined civil society is added to the observation of the shifting public and private parameters relating to identity and knowledge construction identified in chapters 1 and 2, we could ask how all this affects politics in the modern liberal democratic nation-state. The next section will explore the extent to which the proliferation of digital media challenges the normative public/private sphere model for liberal democratic states and what implications this has for individual and collective identities. The public sphere is not only the site for discussing politics in modern democratic societies, it is also where we *learn* about the great ideas of the day and *formulate* our opinions. This knowledge-making function will also be explored in an attempt to link it to the discussions in Chapter 1.

The public sphere

Jürgen Habermas is the theorist who is most commonly associated with the concept of the public sphere. Written in 1962, Habermas's *Structural Transformation of the Public Sphere* was translated from German into English in 1989 and has had a significant impact on intellectual discourse in the Anglophone world ever since. Habermas's theory emanates from his analysis of what have been retrospectively termed as the nascent bourgeois public spheres of the emerging modern nation-states in Europe in the late eighteenth to mid-nineteenth centuries. Put simply, the public sphere is a space which is separate from both the state and the private sphere (family/market) where people are able to discuss rationally and without interference the great public issues of the day: 'By "public sphere" we mean first of all a domain of our social life in which such a thing as public opinion can be formed. Access to the public sphere is open in principle to all citizens' (Habermas 1997/1989: 105).[6]

The type of rational political debate that takes place within the public sphere, however broadly defined, depends on individuals having access to good sources of information. Habermas's concept of publicity was very much dependent on the mass media. The development of public service broadcasting in the twentieth century, mainly, but not exclusively, in western Europe, is an example of the way in which nation-states have attempted to encourage educative forms of media as a means of giving citizens' authoritative information with which to formulate their political opinions.

For Habermas, the first viable public sphere was in Britain – referred to as 'the model case' – at the beginning of the eighteenth century (Habermas 1991/1962: 57–67). Critical to the development of this model public sphere was the role of the press, whose constant battle with the parliaments of the day led inexorably and incrementally to progressive political reform. But it was not only the press that was instrumental in bringing about change, clubs and coffee houses too were venues in which the great issues of the day were debated. Habermas's theory of how 'rational' political discussion can lead to progressive political reform is encapsulated in his idea of 'publicity':

> The critical principle of publicity is an idea which Habermas traces back to Kant's writings on enlightenment: it is the idea that the personal opinions of private individuals could evolve into a public opinion through a process of rational-critical debate which was open to all and free from domination. Habermas wishes to maintain that, despite the decline of the bourgeois public sphere which provided a partial and imperfect realization of this idea, the critical principle of publicity retains its value as a normative ideal, a kind of critical yardstick by means of which the shortcomings of existing institutions can be assessed. The critical principle of publicity is the core concept of a theory of democracy and of democratic will-formation which, at the time of writing *Structural Transformation*, Habermas was only beginning to formulate.
>
> (Thompson 1993: 178–179)

The reference to the 'decline of the bourgeois public sphere' above reflects Habermas's view that the commercialization of the media and – until the recent growth of digital media – its uni-directionality in sending content from powerful media producers to the seemingly passive audience, has had a profoundly negative effect on this normative model. Habermas's critical view of the mass media was undoubtedly influenced by Adorno and Horkheimer (Thompson 1993: 183), and the development of active audience theories have provided convincing counter-arguments to simple depictions of passive readers, viewers and listeners. But whether or not we agree with Habermas's description of the mass media and its audience, the important point here is his intimation that the realization of his ideal public sphere cannot be delivered without a media environment capable of providing an effective knowledge-making function. Indeed, while the focus of his concept of publicity is based on face-to-face communication, in modern societies most communication is mediated. The logic of Habermas's theory is that changes in the wider media environment will affect the development of the principle of publicity, something we should think about in relation to digital media, as the next section will do.

Many theorists of the public sphere are critical of Habermas's conception of it. Nancy Fraser (1992) has stressed the lack of inclusivity in the bourgeois public sphere, especially in relation to Habermas's seeming marginalization of women. Her re-working of the public sphere involves the breaking up of the monolith into myriad public spheres, or 'subaltern counterpublics' (Fraser 1992: 123), to better reflect the diversity of all societies. Others, like Thompson (1993) have questioned whether it is possible (or even desirable) for so-called 'rational' discussion to lead always to consensus, especially where some of the participants are more powerful than the others (Dahlgren 2005: 157). There is a general view however, even among those who are critical of Habermas's conceptual framework, that it is a useful analytical tool to use when engaging with contemporary politics. Thus, theorists like Dahlberg (2007b) who, like Thompson above, are critical of the rational or 'deliberative' model of democracy that Habermas promotes, nonetheless engage with him on his normative terrain. Similarly, Papacharissi's (2010) project to convince us that the public sphere has been eclipsed by 'a private sphere', nonetheless argues that:

> At the present time, however, Habermas' public sphere presents a theoretical model that allows us to discuss the civic *gravitas* of the Internet, contextualize it within the contemporary socio-economic setting, and compare it to that of other media.
>
> (Papacharissi 2010: 113)

Therefore, despite its conceptual shortcomings, the final part of this chapter will use the public sphere as a normative model to analyse the impact of digital media on contemporary politics.

Can digital media reinvigorate contemporary politics?

This final section will attempt to draw together a number of strands from this and the previous two chapters. The first two chapters identified the changing dynamics of the public and private spheres, where the information from which we construct our social and political views is accessed increasingly from private, digital platforms, as opposed to the mass media and memory institutions, and where previously private aspects of our identities are becoming increasingly public. As stated above, the political implications of this shifting dynamic are best explored with the aid of the normative model of the public sphere. This section will build on this by identifying factors specific to digital media which both undermine the normative model of the public sphere and provide opportunities for its reinvigoration, as a means of outlining the key features of politics in the new (digital-media-saturated) public/private environment.

While there were those in the 1990s who responded to the rapid popularization of the Internet with a hope that it could revitalize the public sphere (Poster 2001/1997), most contemporary academic writing takes a more pessimistic view. Much of this intellectual pessimism derives from Manuel Castells's monumental trilogy on the *Information Age,* which was published in the mid- to late 1990s and updated twice since. There, Castells (2010b: 11) argues that the global network has undermined civil society so much that it is now largely irrelevant to meaning-making. This is a problem because while, as Downey and Fenton (2003) argue perceptively, we should not conflate the two concepts, successful public spheres do tend to rest on progressive civil societies. Even some of the more optimistic accounts of the so-called new politics are based on the premise that the 'old politics' (including, of course inconvenient notions like the public sphere) has been superseded (Papacharissi 2010; see also White's (2012) discussion on the politics of *Wired* magazine writers). For this reason, this section will answer its eponymous question by first addressing the view of the sceptics. But, as I have argued elsewhere (White 2012), it is a mistake to view digital media as almost solely good or almost solely bad, and so this scepticism will be followed by a more nuanced account with which to end the chapter.

Digital media and the undermining of the public sphere

The principle of publicity, outlined above, stressed the importance of the formulation of critical knowledge by citizens as a foundation of democracy. Habermas argued that the mass media as a facilitator of the construction of publicity is flawed, as its structure does not enable the audience to develop critical thinking (Cavanagh 2007: 63). In particular, the role of public relations in the modern media bedevils attempts to arrive at the 'truth' (Boeder 2005). Cavanagh (2007) argues that communication theorists place too much emphasis on the role of the media when they are discussing the public sphere and, in this sense, we could argue that people can turn to other sources of information. As Chapter 1 illustrated, these could include libraries, archives and museums, and the influence of these institutions within the public sphere has certainly been under-researched. But, as that chapter also showed, the increasing digitization of sources from those institutions – the shifting of information from public institutions to private platforms – runs the risk of endangering the type of academic conventions like provenance and classification which make these information sources credible. Cavanagh's (2007: 63) argument that the mass media undermines critical enquiry because 'there is no Archimedean position outside of the system from which to reflect upon it, no gap between the institutions of one domain and those of another' is even more applicable to digital media. Commenting on the role of the Internet in contemporary politics, Habermas (2006) himself argues that: 'This [the Internet] focuses the attention of an anonymous and dispersed public on *select* topics and information, allowing citizens to concentrate on the *same* critically filtered issues and journalistic pieces at any given time' [emphasis in the original]. Here, Habermas is supporting the argument of those cited in the first two chapters who argue that it is not so much the problem that the Internet is providing a surfeit of inaccuracies and falsehoods but that it is exposing us to an ever narrower range of information.

But there is a more serious challenge to the public sphere in relation to its knowledge-making function. The type of knowledge that the traditional public sphere generated, be it through the mass media or the memory institutions, took place within a national context. National libraries, archives and museums gave citizens knowledge of their national culture. In the mass media, public service broadcasting regimes (mainly in western Europe) also have a remit to educate and inform citizens about the particular national culture to which they belong, and even the newspaper industry tends to focus on national news. While these national public spheres could be restrictive in their knowledge-making function, the globalization of media best exemplified by the World Wide Web will undoubtedly alter not only the way in which we learn about the world around us, but also the way in which we construct our political views. While this change has positive elements – it can make us open to new ideas beyond our own culture and can expose immigrants to more content from their countries of origin – there are undoubtedly negative aspects too, in particular the sense in which it fragments the public sphere.

Sunstein's (2007) argument in the previous chapters provided an informative amalgamation of discussions about identity and knowledge construction in this new global context. In his reading of contemporary trends, the World Wide Web is indeed being fractured along tribal lines or what we might refer less pejoratively to as communities of interest. That these are essentially communities of like-minded people creates epistemological, as well as communal, divisions. In other words, communal identities not only divide people along the lines of the particular identities that they foreground, but they also produce subjects whose meaning-making is tied to those communal identities.

If the way in which we construct knowledge is dependent on our tribal identity, this undermines the universalism on which rational discussion within the public sphere is based. He concludes that societies in which Internet users retreat to silos of like-minded people encourage polarization and, *in extremis*, political violence. This could account for the contemporary resurgence of the far-right and of radical Islam. In relation to the former, Downey and Fenton (2003), albeit writing over a decade ago, refer to the sophistication and large quantity of far-right websites. Dahlberg's (2007b: 136–137) analysis of one of these websites highlighted the dearth of dissenting voices, a classic trope of the tribal website. Are these activities confined online or do they encourage polarization in wider society? The rise of a global, networked Islamic fundamentalism would suggest that the latter is more likely (Castells 2010b). The recent case of the home-grown suspects in the bombing of the 2013 Boston Marathon focused on the Internet activity of these otherwise seemingly well-integrated American Muslims. Acquaintances of surviving suspect Dzhokhar Tsarnaev were surprised when they saw his first court appearance:

> 'He never had that accent', said a friend from the school, declining to give his name as he spoke to reporters outside the courtroom. He had come to see Dzhokhar in court along with other friends from the suspect's high school wrestling team. He insisted that not only had Dzhokhar's voice changed but that his demeanour was also different to what he had known. 'He didn't fidget,' he said.
>
> (Kumar 2013)

While it would be simplistic to argue that exposure of the suspects to jihadist websites was the sole factor in their alienation from American society, these alternative sources of information outside what constitutes the national public sphere are becoming ever more potent. And, as the more positive example of Autumn Edows in Chapter 2 illustrated, frequent exposure to them can alter individuals' demeanours in the real world. There might be other reasons for Tsarnaev's behaviour, but one would be hard pressed to explain why a teenager, whose only contact since childhood with the region of the world where he was born appeared to be through a Russian-language website, behaved in this way without discussing the role of the Internet in the supersession of national public spheres. This tale cautions us against assuming that the World Wide Web provides opportunities mainly for those of a more 'progressive' bent.

In her book *A Private Sphere*, Papacharissi (2010) argues that we must accept that most people nowadays practice politics from private platforms. As reported in the previous chapter, this assertion is based on the assumption that people can retain a strong, private sense of self as they navigate online. While never fully embracing the ideas of Bauman (2000) and Giddens (1991), there are many allusions in Papacharissi's book to liquid and self-reflexive identities. And the networked self that she advocates does assume a great degree of individual autonomy. But this autonomy can encourage an individualistic, ego-centred form of politics that is more concerned with maintaining one's network of contacts rather than critiquing the political economy of that same network (Fenton 2012). Fenton's (2012: 142) argument that the networked self tends to fetishize speech at the expense of critique suggests that it fails the test of publicity, namely, being discussion that does not lead to meaningful political reform. We also cannot ignore the commercial element of networked identities, something that limits attempts to craft progressive political projects online (see previous chapter and Fenton (2012)). Perhaps partly because of the networked self's shortcomings in relation to political critique, as was identified throughout the last few chapters, identities anchored in traditional cultural resources have proved to be resistant even in an online environment. And, being largely communal in nature, these identities encourage solidarity with other members of the tribe rather than more ephemeral attachments. A strong private sense of the self (manifest in online anonymity) insofar as it relates to online identity has also been undermined, through the need to establish trust both for commercial reasons and for social networking. Castells (2010b: 11) sums up well the argument against reflexive forms of identities:

> reflexive life planning [within the global network] becomes impossible, except for the elite inhabiting the timeless space of flows of global networks and their ancillary locales. And the building of intimacy on the basis of *trust* requires a redefinition of identity fully autonomous *vis a vis* the networking logic of dominant institutions and organizations [my emphasis].

Where Papacharissi's (2010) views have more resonance is in her emphasis on the dramaturgical nature of online identity. But while she describes blogging as 'a new narcissism' (Papacharissi 2010: 144), for her this does not necessarily connote negativity. Instead, this enables the blogger to critically self-reflect on his/her own politics. While this does not contribute to the health of the public sphere (which, remember, Papacharissi rejects as a normative for political interaction in contemporary digitally saturated societies), she argues that it does disseminate views alternative to the values of mainstream society (Papacharissi 2010: 149). We can also tie narcissism and the dramaturgical nature of online identity to Habermas's concept of publicity: if you want to influence political debate in the wider world, then people must first notice you; ergo, as the previous chapter noted, you must have a persona which enables you to get noticed.

But while a dose of the dramaturgical can help marginalized activists get their message across, it is more problematic when applied to mainstream electoral politics. Castells (2010b) argues that while the media, digital or otherwise, does not directly influence the voters, electoral politics has, to a large extent, submitted to the logic of the medium through which it is framed. Ever since the 1960 US election, where Kennedy's television persona was believed to have won him the race to the White House, the media has been increasingly successful in promoting electoral politics as a form of entertainment. Image is therefore important, as well as the promotion of 'stars'. The proliferation of digital media has exacerbated this, with image (in the sense of identity) becoming ever more difficult to manage, while imagery (literal translation) becomes ubiquitous. There are two main consequences stemming from this trend. While the media in and of itself cannot deliver individual candidates electoral success, it is virtually impossible for a candidate with less than a rudimentary grasp of how to use it to win national elections (Castells 2010b: 375). Also, politicians' personalities become central to their political programmes, thus empowering the medium (namely digital media) that has the capacity to find and disseminate embarrassing information that might undermine a carefully constructed persona (Castells 2010b: 391–402). The political paralysis in Washington, DC in the mid-1990s induced by the Drudge Report website's revelations about President Bill Clinton's sexual misdemeanours has ushered in an era defined by what Castells (2010b: 368) has termed 'informational politics'. While many of the issues in contemporary informational politics are serious (here, one thinks of Wikileaks and the scandal of British MPs' expenses in the UK) rather than simply personal indiscretions, the oxygen that digital media gives them renders them more newsworthy than other issues that cannot be so easily packaged online. An example of this was the aforementioned British MPs' expenses story at the height of the global financial crisis in 2009, which undermined the integrity of elected politicians while conveniently diverting attention from the role of financiers and bankers in precipitating a global downturn. Similarly, it is questionable whether the revelations of Wikileaks and other organizations pass the test of publicity: it is not clear that they have impacted on public policy in the USA in anything other than the most cursory way (White 2012).

Counter-publics

The section above paints a rather dark picture of the influence of digital media on contemporary politics. But, as promised, this chapter will finish on a brighter note, focusing on one area where digital media has the potential to aid progressive political projects. As noted earlier, some theorists are critical of the deliberative nature of Habermas's public sphere normative model (Dahlberg 2007b; Dahlgren 2005; Thompson 1993). That is to say, rather than try to seek consensus, they think that politics should be more closely aligned to the agonistic than the deliberative. For some like Papacharissi (2010: 161), who uses the work of Chantal Mouffe (2005) to support a political model which is both agonistic and democratic, this invalidates the public sphere normative. But others have attempted to incorporate this antagonism into a normative model of the public sphere (Dahlberg 2007a, 2007b; Downey 2007; Downey and Fenton 2003). How, then, might we construct a more optimistic normative model for online politics?

Nancy Fraser's (1992) criticism of Habermas's apparently monolithic conception of the public sphere opens up an avenue for an alternative conception based on 'counter-publics'. This position involves accepting the view held by Mouffe (2005) and Papacharissi (2010) that politics is inherently agonistic, but rejecting their conclusion that this ineluctably leads to an abandonment of the public sphere as a normative model for contemporary politics. In seeing the public sphere as the reflection of the interests of dominant individuals and institutions in society, counter-publics are seen as a vital means of empowering those whose values are not reflected in it. An important aspect of this theory is that the dominant public sphere should be receptive to reform, thus obviating the need for its rejection as a normative ideal or, *in extremis*, outright revolution. Downey (2007: 118–119) argues persuasively that we should think of the public sphere as being constantly in flux, rather than a monolithic entity. For Downey (2007: 118–119), this flux can be explained as the consequence of the disjuncture between the normative claims of liberal universalism and the actuality of domination of the public sphere by the most powerful elements of any given society. This capacity for 'immanent critique, critique from within or from the perspective of liberal universalism' (Downey 2007: 119) was recognized by Habermas (1996: 374; cited in Downey 2007: 119) himself:

> In the course of the nineteenth and twentieth centuries, the universalist discourses of the bourgeois public sphere could no longer immunize themselves against a critique from within.

As well as this, elite actors' powers wax and wane, and when they do counter-publics have the opportunity to make decisive interventions (Downey 2007: 119). An example of how this how this might work in actuality can be demonstrated using the global financial crisis as an example. Until around 2007, neo-liberalism as the public sphere's dominant economic doctrine faced little serious challenge. But the global financial crisis undermined the power of both the financial elites and the powerful politicians who supported them. This has provided an opportunity for counter-publics based on opposition to global neo-liberalism to develop critiques that have had some impact on the wider public – here, one thinks of organizations like Occupy, a global organization devoted to combatting global inequalities.

But if we think about this in terms of Habermas's concept of publicity, how do counter-publics find their voice in the competitive global marketplace of ideas that digital media facilitates? This can best be achieved by challenging the dominant powers at the level of discourse. This can only be done if different counter-publics work together to find a common language. While, as has been noted throughout, digital media often encourages insularity as interest groups create their own hermetically sealed silos, failure to construct a wider discourse will make radical groups 'vulnerable to being isolated, assimilated, or reduced to interest group politics' (Dahlberg 2007b: 137). Counter-publics have had some success in recent years in creating a shared critical discourse against global neo-liberalism, the caveat being that attempts to establish broad churches inevitably often result in a lack of clarity in offering alternative models of governance (White 2012).

This bent towards cooperation with others is crucial not only as a means of challenging dominant discourse but also as a strategy to overcome the imbalance in power between a largely unified hegemonic *bloc* and disparate sites of resistance:

> Unless powerful efforts at alliances are made – and such efforts have been made successfully, especially in the area of the environment, globalization and ecology – the oppositional energy of individual groups and subcultures is more often neutralized in the marketplace

of multicultural pluralism, or polarized in a reductive competition of victimizations.

(Downey and Fenton 2003: 194)

Though the overall normative model in which these formations take place is agonistic, the act of formation itself is best done deliberatively (Dahlberg 2007b: 142).

The deliberative formation of counter-public spheres is done to avoid some of the problems of fragmentation outlined above and, earlier, by Cass Sunstein. Parekh's earlier argument that tribes can sometimes enable groups to build internal self-confidence needs to be revisited here. Both Dahlberg (2007a: 833) and Downey and Fenton (2003: 190) quote Cass Sunstein's (2001: 75) own admission that in cases like the nineteenth-century anti-slavery movement, building an enclave or haven was helpful rather than detrimental, not only to the cause itself but to wider society too. There is also an empirical challenge to Sunstein's echo-chamber thesis in the highlighting of research that suggests that Internet users *do* expose themselves to a wide range of ideas and debate with people whose opinions they do not share (Dahlberg 2007a: 830). What is required to ensure that these enclaves or havens do not become self-serving or, worse still, reactionary, is for them to incorporate the type of immanent self-critique that Downey advocated for the dominant public sphere (Dahlberg 2007b: 139, 141) (Downey and Fenton 2003: 194). (For this reason, Dahlberg (2007b: 139) argues that reactionary counter-publics are problematic in their failure to contribute to an agonist politics with which to challenge mainstream discourse.)

Surprisingly, this progressive political project has a role for the dramaturgical, as suggested by Papacharissi (2010) earlier. For it is the dramaturgical intervention that is most likely to succeed at those points where the disjuncture between the ideals and actuality of the public sphere create openings (Downey 2007: 119). This explains the increasingly carnivalesque nature of modern political protests, whose spectacles can be easily recorded and disseminated worldwide by digital media technologies (see Lovink's (2011: 178) discussion on Wikileaks use of the 'spectacle' in getting its message across).

Conclusion to Part I

The last few pages have illustrated the continuing importance of the liberal democratic nation-state and have attempted to convince the reader that the public sphere is still the most credible normative model for conceptualizing political activity within it. How, though, does this relate to the themes outlined in the first two chapters?

Chapter 1 was premised on the notion that identifiably separate public and private spheres organize politics in modern liberal democratic states. While the increasing privatization of many formerly public institutions, the continuing movement of academic material through a process of digitization into commercial organizations, and the move away from notions of privacy that social networking and scandal-based politics represent, suggest that the boundaries of the public and private spheres are blurred, if not over-lapping, the continuing importance of this normative model means that it is too soon to condemn these spheres to history. Indeed, the legal challenges to which the Google Books project has been subject (Woollacott 2013) suggest that we should be very careful about writing the obituary of those components of public knowledge – provenance, classification and the general idea of information held in the commons – seemingly imperilled by the rise of digital media.[7] Instead, it may well be the case that digital media, like other media before it, has to find a way of incorporating these into its working practices rather than trying to eliminate them.

In terms of some of the political developments that have been formulated in this chapter, it seems that the need for credible information is greater than ever before, which is not always easy in a digital media environment. Downey's (2007: 118–119) argument that the public sphere and, by extension, counter-publics are constantly in flux is similar to what this author has argued elsewhere (White 2012). Indeed, the latter's argument emphasized the speed with which these fluctuations occur, which means that activists must get used to responding quickly to this shifting landscape. How do they do this? Through a process of constant reinvention of their tactics and, sometimes, of their arguments. This requires good, credible information sources on which both to craft their arguments in a rapidly shifting political environment and to achieve the self-reflection that is *de rigeur* for successful counter-publics. Finding these credible information sources will require a reversal of the trend towards the ever greater privatization of search engines and knowledge platforms. With this in mind, the announcement in April 2012 by Harvard Librarian Robert Darnton of the development of the not-for-profit Digital Public Library of America is a welcome initiative (Flood 2012).

As explained in Chapter 2, identity has become more performative or dramaturgical as a result of the need to be noticed online. Paradoxically and simultaneously, there has been a strengthening of the influence of traditional cultural resources in the construction of identity, in contradistinction to the views of late-modern identity theorists. These two trends have undoubtedly been exacerbated by the proliferation of global digital media. This has both negative and positive consequences, giving a platform to extreme nationalism and religious chauvinism and serving as a means by which their adherents can link up with like-minded people in different parts of the world. Dramaturgical identities have also exacerbated the trend towards celebrity or informational politics. The undermining of people's privacy that digital media encourages and informational politics' seeming rejection of the public sphere ethos of respect for all participants makes contributors vulnerable (as I write this the latest in a long line of depressingly common threats against politically active women using Twitter is headline news in the UK (*BBC News* 2013a)). In a more positive vein, progressive activists can use digital media to make dramaturgical interventions in the public sphere that can meet Habermas's test of publicity. Chapter 7 will give some examples of these. The strengthening of identities inscribed with traditional cultural resources can provide the glue for the solidarity needed to challenge the official public sphere (as the author has made clear elsewhere (White 2012), it is not clear how a progressive politics can be fashioned out of liquid or postmodern identities).

Given its focus on the theoretical, this chapter has eschewed lengthy examples of political campaigns which utilize digital media. This will largely be addressed in Chapter 7's discussion of new social movements. It is also important to note that this theory applies mainly to politics in liberal democratic societies. It would therefore be unwise to apply these theories universally, especially to societies which are not democratic or not fully industrialized. Chapter 9, therefore, will focus on the differing ways in which digital media is used in the politics of various developing nations. A lacuna in this discussion about politics is the omission of a serious discussion about economics. Rectifying this will be the role of Part II of this book.

Part II
The Digital Economy

Introduction to Part II

This is the first of two chapters which focus on the impact of digital media on both the global economy and national economic strategies. The previous chapter was replete with references to the global economy and the network society and the reader might have lamented the seeming unwillingness of the author to define and discuss these issues in greater detail. The reader might also wonder why concepts so central to our understanding of modern politics hovered in the background rather than at the forefront of the discussion. The answer to this relates to the author's concern that many discussions about the network society in particular veer dangerously near to a form of economic determinism. Similar to Marxist accounts of politics, those who write about the network or information society often give the impression that once we understand the base (world economy) then the complexity of the super-structure (national cultures and societies) can easily be explained. So, if I were to explain to the reader my reasons for veiling global economics in the first section, they would essentially amount to me wanting to highlight what I believe is the continuing importance of the nation-state as the pre-eminent vassal in which the most pertinent political issues of the day are discussed. That is not to say that the changes that will be outlined in this section have not been influential in re-shaping the nation-state – the previous chapter highlighted some of the challenges that the liberal democratic state has had to engage with in order to retain its pre-eminence. But it is important to discuss these changes within the context of the politics of the international system of nation-states – hence the reason why the political discussion in this book preceded the economic one.

Having said all that, it is, of course, neither easy nor desirable to disentangle the political and the economic. Indeed, as Hesmondhalgh (2007: 87) has noted, many writers and politicians use interchangeably the terms 'information society' and 'information economy'. For this reason, the reader will be introduced to some political arguments, pertaining as they do to contemporary global economics, which he/she did not encounter in the first part of this book. But this entangling of economics and politics is intended to be an antidote against both the tendency to view economics through the positivist prism of its many academic practitioners and the more pernicious tendency of many neo-liberals to persuade us that the present global economic framework is 'natural' or 'commonsensical'; its critics thus being 'irrational' and 'political'. We are all political; the most political among us are those who deny their own politics. In this tenor, Part II of this book will analyse the political economy of global digital media as a means of both adding to and complementing the arguments in Part I.

Although, like all the other chapters, the fourth will provide a critical account of the main arguments, its discussion of them will emphasize their most convincing points. It will set out the strongest arguments in favour of the proposition that, whether we use the prefix post-industrial, network, digital or information, we now live in societies which are powered by the production and selling of intangible or symbolic goods. Daniel Bell's (1973) theory of post-industrialism, David Harvey's (1990) study of the postmodern economy and Manuel Castells's (2010a) network society thesis will primarily be used to explain the rise of the digital economy. Alongside analysis of these general theories will be a discussion of nation-states' responses to these changes in the global economy, with an emphasis on the most technologically developed of those states. This will involve discussion of national strategies for developing the digital and creative economies within various jurisdictions. The chapter will end by introducing some of the tensions that have arisen from the pursuance of such strategies, and these will be more fully explored in Chapter 5. The latter chapter will focus primarily on the role of digital media in the global financial crisis of 2008. It will consider whether this crisis might be the result of deep-rooted and irreversible changes in the world's economy, rather than merely being a cyclical downturn to which global capitalism is prone. In particular, the role of digital media both in encouraging automation and offshoring, and in fuelling the seemingly out-of-control global financial system, will be scrutinized.

4
The Digital Economy and the Creative Industries

Introduction

Debates about the seeming shift in the economic structure of societies away from agriculture and manufacturing, and towards service industries, have been taking place, in the USA at least, since the early 1970s. Daniel Bell's 1973 book *The Coming of Post-Industrial Society* is widely regarded as the seminal text, but it was preceded and succeeded by others, including Machlup (1962), Drucker (1992/1968), Touraine (1969), Toffler (1970), Porat (1977), Toffler (1980) and Naisbitt (1982) (Cavanagh 2007: 156–157; Chadwick 2006: 29–30; Hesmondhalgh 2007: 87–88; Lister *et al.* 2003: 192–197; Miller 2011: 49–54; Qiu 2013: 110–113). While Miller (2011: 52–53) is keen to make a subtle distinction between the concepts of post-industrialism and the information society, the differences between the two are more temporal than philosophical in orientation. This is exemplified in the development of Bell's own work, whose post-industrialism morphed into the 'information society' thesis over time (Miller 2011: 52–53). Post-industrialism described a period in the 1970s where service industries were becoming more prominent without there being the technology available for them to fully exploit the new opportunities for growth. Later theories of post-industrialism in the 1990s had the advantage of being able to incorporate the development and use of advanced forms of digital communication as well as the rise of neo-liberalism into their analyses. The most prominent of these analyses was Manuel Castells' magisterial trilogy on the 'network society', published in the mid-1990s and re-issued twice since then. This chapter, therefore, will begin with a discussion centred on the work of Daniel Bell and proceed, through an intervening discussion on the work of David Harvey, to Manuel Castells' network society thesis as a means of demonstrating how the concept of post-industrialism has been developed and incorporated into a theory of global economics based on advanced and always-available global communication networks, before ending with a look at a range of arguments which identify the opportunities and challenges afforded by the digital economy.

The post-industrial society

Daniel Bell's theory of the post-industrial society was an attempt to explain the underlying reasons for the changes that were taking place in western societies in the early 1970s. Sociological both in its drawing on statistics from social surveys and its identification of long-term trends which supposedly govern human behaviour, post-industrialism comprised the following elements: 'in the economic sector, it is a shift from manufacturing to services; in technology it is the rise of new technical elites and the advent of a new principle of stratification' (Bell 1973: 487; quoted in Qiu 2013: 110). It was explicitly teleological in its description of a society advancing from the pre-industrial to the industrial, and thence from the industrial to the post-industrial. In this new society, the most important factor in the workplace was workers' theoretical knowledge. This knowledge was developed society-wide through the increase of research and development in universities, scientific institutions and corporations. But each worker was also expected to improve his/her own work-based knowledge, which was both a reflection and reinforcement of the increasing individualism of society. Organizations were encouraged to become less hierarchically structured in order to incorporate these changes in their workforce (Turner 2006: 242). It should be noted that Bell, who was highly critical of the counterculture movement in the USA, which could be said to have contributed to this new zeitgeist (Turner 2006: 32, 245), was concerned about the undermining of authority that the post-industrial society brought in its wake (Harvey 1990: 353).

There was some empirical support for Bell's thesis, in the USA at least (which was the country his study was predominantly concerned with), in the statistical observation of long-term occupational trends. From 1870 to 1978, agricultural employment fell from 47 per cent of the national total to 3 per cent, while, from 1920 to 1970, workers in the extractive industries (agriculture and mining) dropped from 28.9 to 4.5 per cent (Castells 2010a: 304). In contrast, the period from 1870 to 1978 saw employment in the service industries rise from 26 to 60 per cent (Fuchs 1980; cited in Miller 2011: 50); employment in distributive, producer, social and personal services all rose significantly from the years 1920 to 1970 (Castells 2010a: 304). While the post-industrial thesis has been criticized for describing as a rupture something which has been much longer in gestation (Miller 2011: 51) and for exaggerating the decline of manufacturing industries (Chang 2011: 88–101), the early 1970s do seem to mark the start of a new type of economy, more informational and global in character.

Harvey's postmodern economy

There are two main factors which have given post-industrialism more credibility since the publication of Bell's book: one economic; one technological. Coincidentally, Bell's book was published around the end of the long post-war economic boom in the western world. In David Harvey's authoritative account of the rise of the postmodern economy, 1973 is marked as the year in which the shift from Fordism (a common term for the type of economy based on the mass production and standardization of manufacturing products best exemplified in Ford's motor car factories) to what he terms an economic regime of 'flexible accumulation' became evident (Harvey 1990: 141–172). While Harvey's essentially Marxist account views the rise of neo-liberalism – exemplified by the coming to power in the USA and the UK of Ronald Reagan and Margaret Thatcher in 1981 and 1979, respectively – as the *result* of longer-term trends which he relates to the contradictions inherent in global capitalism rather than the *cause* of a new economic direction, it is hard to deny that the dominance of that ideology continues to impact on the make-up of global economics. Harvey's thesis is that all capitalist economies are inherently prone to over-accumulation – caused by the tension between the unending need for growth and the consequences of that growth – which is 'indicated by idle productive capacity, a glut of commodities and an excess of inventories, surplus money capital (perhaps held as hoards), and high unemployment' (Harvey 1990: 181). The periodic crises that capitalism faces (the so-called boom-and-bust phases) are therefore inevitable. Harvey (1990: 181–184) explains many ways in which capitalism tries to reinvigorate itself after a crisis but what is of most relevance here is its precise manifestation in the early 1970s, which is detailed in the following.

Harvey (1990: 141) argues that the first signs of the economic crisis of the early 1970s were detectable during the 1965–1973 period when internal markets in the advanced capitalist nations, most notably in western Europe and Japan, became saturated. Like industrial conditions within Ford assembly plants, Fordism in macrocosm was based on full employment and decent wages as a means of sustaining the demand for new products. What to do, then, when demand stagnates – when each family has bought virtually all the household goods it needs? In a capitalist system that is predicated on perpetual growth, the answer is to seek new markets. Also, though capitalism is always implicated in driving down production costs to increase profits (Harvey 1990: 180), this becomes even more necessary in times of economic crisis. Thus, regions outside the advanced capitalist zones became targeted both as new markets for western goods but also as places where production itself could be carried out with much lower labour costs than in the host countries of large corporations. In this new regime of 'flexible accumulation', not only was the cost of labour cut, but the labour practices in typically non-unionized environments overseas were fed back into those heavily unionized advanced capitalist nations (Harvey 1990: 147). Over time, this created a global workforce that was supposedly more 'flexible' – for corporations certainly but not always for workers. It also accentuated moves towards post-industrialism in the advanced capitalist nations, manifested mainly in the growth of the service and (to use a term coined by the UK's New Labour government in the late 1990s) creative industries. Why was this so? Harvey (1990: 156–157) puts it down partly to the accelerating turnover time of capital as a response to the economic crisis. Fordist production had created products that lasted for years. This was fine during a period when economies still had a lot of slack, but not so good when they were near to full capacity, resulting in a slowing demand for basic household goods. The key, then, was to cut the shelf-life of products so that people would continue to buy things as regularly as before, even if they did not really need them. Survival in the marketplace thus became dependent on innovation and the creating of niche products. In sectors like computer software and gaming it is easier to convince people that they should upgrade every 18 months than it is in industries that sell more functional products (Harvey 1990: 156). Thus, these types of industries became much more important to the functioning of the new economy, especially in the most advanced capitalist nations. But even traditional manufacturing was not immune to these changes, as design and advertising transformed even the most basic appliances. Manufacturers try to convince us that we need new products through frequent re-design and with clever advertising and marketing. Take the motor car as an example. Henry Ford's maxim that you can have any car you want as long as it's black was perfect for a market where very few people owned cars and useful in delivering a vehicle that would last for a very long time. Modern car manufacturers, though, constantly re-design their models as a means of trying to get us to part with our money every few years. Advertising too is deployed essentially to create desire for something that we really do not need. This is taken to its logical extreme with branding, where a product's price is determined, sometimes in a massively disproportionate way by its brand. (We can see this *in extremis* with the cost of items produced by the world's most successful fashion labels).

But these economic changes cannot be explained without also thinking about the growth of technology, particularly that of an informational kind. Harvey (1990: 159–160) himself emphasizes the importance of information to this latest phase of capitalism, both as an end in itself – in many of the new industries which are informational in nature and policed with strict intellectual property regimes – and in the need for the modern corporation to respond rapidly to changes in consumer tastes. Because the modern corporation has branches in many different countries, the need to send information quickly over long distances requires more advanced forms of communication. While it has long been the case that corporations have had overseas interests and the means by which to communicate with their bases back home, the scope and speed of this type of communication has expanded exponentially since the 1970s; an account of which will proceed below.

Exponential growth in computing power

In 1965, computer engineer Gordon Moore identified something about the increasing capacity of computer chips which became incorporated into a theory which bears his name (Kelly 2010: 161). Moore's Law stipulates that computer processors double in power around every 18 months (Castells 2010a: 39). Few would doubt that Moore's Law has proved to be correct, as computer processing power has grown rapidly since then. For some authors, because this growth is exponential rather than incremental, this involves a qualitative change in computing power rather than merely a quantitative one (Ford 2009; Kelly 2010). This position is taken to its logical extreme through the work of Raymond Kurzweil, who argues that we are nearing the 'singularity', the point at which technological growth becomes so rapid that machines will be smarter than human beings (Ford 2009: 100–101). While, as the work of Seidensticker (2006) points out, increased computing power is not matched by commensurate efficiency gains (students cannot double the speed with which they type their essays every 18 months!), one does not need to indulge the futurism of Kurzweil to acknowledge that technological growth has really taken off since the 1970s. Around the time that Daniel Bell was writing about the post-industrial society, the microprocessor was being invented by Ted Hoff (in 1971). By 1977 Apple had built the first complete personal computer (PC) (Lister *et al.* 2003: 196) and, throughout the 1980s, there was a steady rise in the ownership of PCs. Alongside this was a commensurate growth in networked communication, with the Internet advancing from a system with four hosts/servers at its inception in 1969 to one with 19,540,000 hosts by 1997 (Lister *et al.* 2003: 165). The effect that this had on the development of global business was plain to see:

> The head office of a large enterprise operating within one national territory during the 1970s could have run on the basis of paper-based communications, memos, postal services, phone calls and the occasional telex. By the end of the 1980s such an enterprise, if it survived at all, would in all likelihood have opened branches in other countries to support its globalised marketing, and would need at the very least a constant stream of fax communications and improved telecommunications of every kind. It would increasingly rely upon telephone lines and network-based communication systems, including commercial satellites, which had begun to be deployed in the early 1970s.
>
> (Lister *et al.* 2003: 197)

While Harvey's account could be said to represent a kind of economic determinism that underplays the extent to which culture can operate according to its own logic (Castells 2010a: 492–493; Hesmondhalgh 2007: 102), an over-emphasis on technological developments alone could be viewed as similarly deterministic. Instead, and in keeping with the tenor of the last chapter, it is to the political that we should turn for a comprehensive analysis of these changes. That is not to say that the scope of political decision-making has not been heavily circumscribed by these economic and technological developments, but that those developments cannot be isolated from wider politics. While it sometimes seems a compelling argument to make that the politics of the day merely reflect the prevailing economic and technological systems, one example will serve to demonstrate the falsity of this position. One early formulation of a computerized society was Jacques Ellul's *The Technological Society*, written in French in 1954 and translated into English a decade later. Ellul employed a technological determinism which argued that once *technique* became embedded in society then the political and the economic would have to bow to its logic. But what kind of society would emerge? Ellul (1964/1954: 184) argued that *technique* favoured central planning and thus societies would become underpinned by planned, rather than market, economies. It could be argued that Ellul was unable to think beyond the parameters of the political situation as it was in 1954, when even in western democracies centrally planned programmes like rationing were fresh in the mind. Ellul's pronouncements seem very dated now, which is probably why he is not even mentioned in the many contemporary accounts of the post-industrial society, including Castells (2010a), Harvey (1990) and Kelly's (2010). But in a sense he is a spectre at the feast, a reminder of the folly of automatically assuming that the prevailing economic and political climate is an inevitable manifestation of the present state of development of technology and/or capitalism.

Politics and the post-industrial society

If we assume that there has been a move from industrialism to post-industrialism and thence to the information society, then this has been as much a political project as anything else. While computerization has evolved with amazing speed in the last few decades, it could not have had the impact it has had on wider politics and society were it not for the development of networked communication. The eclipsing of the publicly oriented Internet with the more commodified World Wide Web took place during the US presidency of Bill Clinton (1993–2001). In a sense, this was not coincidental, as his vice-president Al Gore had been one of the first senators to champion the 'information superhighway' (Mosco 2005: 38–39). But his vision of an Internet in which the US government would defend the public interest against the commercial organizations that would also be part of its platform did not materialize when he was in office (McChesney 2013: 116–117). Instead, by March 1994 Gore had outlined a vision of the Internet at an International Telecommunications Union (ITU conference) in Buenos Aires which 'propose[d] that private investment and competition be the foundations for the development of the GII [Global Information Infrastructure]' (Gore 1994; quoted in Schiller 2001/1995: 159–160). The Internet was privatized shortly afterwards when the National Science Foundation Network (NSFNet) handed over the backbone of the Internet to private interests (McChesney 2013: 104). All this culminated in the 1996 Telecommunications Act, the first such act since 1934, which instituted widespread deregulation in the telecommunications sector, including digital media, especially in relation to ownership (Hesmondhalgh 2007: 117).

While running the risk of excessive repetition, it must be stressed how dependent these changes were on the confluence of a number of political events and ideas. While nominally centre-left, the Clinton administration was committed to deregulation in many sectors of the economy, including, as will be detailed later, finance. There was also intensive lobbying by the corporate world and a Congress that at this stage of the Clinton administration prioritized factional fighting over detailed scrutiny of policy (McChesney 2013: 106–107). But perhaps the main reason was the political culture of the USA, which, especially at this time, was hostile to federal government intervention, even where that might curb the power of large corporations. That is why the main opposition to Internet-related legislation at the time was the supposed threat to free speech posed by the Communications Decency Act of 1996 (McChesney 2013: 105). Clinton's own justification for the Telecommunications Act was that it would increase economic competition on the Internet is telling in its abandoning of any vestige of pretence that the public interest should play any sort of role in its future development (Branch 2009: 349–350). At the time, the USA was the world's sole superpower and was effectively governing the Internet during this period of time, and was thus able to influence worldwide policy. Indeed, western Europe nations, which had an even stronger public interest model of communication underwent similar privatizations and deregulation in the 1980s and 1990s (Hesmondhalgh 2007: 119–121). This was reinforced by the creation of the World Trade Organization in the 1990s, under whose aegis were the telecommunications policies of individual nation-states.

Sociologist Manuel Castells has made the most famous attempt to critique these changes, drawing on Harvey, Daniel Bell and one of his mentors, Alain Touraine, to build a theory based on the 'network society'. Like Harvey (1990) and others (Bauman 2000; Giddens 1991), Castells is concerned primarily with the impact of these economic changes on our perceptions of time and space. The next section will explore this in more detail and discuss the implications of Castells's theory of the network society for the financial industries, other existing industries and the growth of the creative industries.

Time and space in the network society

The network society, a theory developed by Castells in a trilogy of books collectively entitled *The Information Age* first published in 1996, 1997 and 1998 and updated twice since, is primarily focused on economics. While Castells broadly agrees with David Harvey's (1990) approach, he stresses: 'that he [Harvey] gives to capitalist logic more responsibility than it deserves for current processes of cultural transformation...' (Castells 2010a: 492–493). Nonetheless, as was illustrated in the previous chapter, Castells (2010a) implies a kind of soft determinism in his argument that politics in contemporary society is mainly shaped by the logic of the network. In particular, he argues that the network society has had a profound impact on global economics, the subject of this section. In order to understand Castells's arguments about networked economics, we must first, though, consider his ideas about the role of time and space in the network society.

David Harvey (1990) had earlier developed a theory of postmodernity based on 'time-space compression'. Improved telecommunications and the increasingly rapid turnover of capital had drastically reduced decision-making times and pressurized industries into constantly reinventing their products. These new forms of communication have enabled the conquering of space. Ironically, this latter development has enhanced rather than diminished the importance of 'place', as regions, cities and districts highlight their distinctiveness in order to attract international capital (Harvey 1990: 293–294). However, this new configuration of time and space benefits those who are not tied to a particular place – multinational corporations, for instance, can threaten to move their operations to places which offer them a better business climate, regardless of the damage that that might do to the region or country in which they are presently located.

Another prominent social theorist who has mused on the time-space nexus in late modernity is Zygmunt Bauman. While Bauman's formulation differs from Harvey's in its grounding in a world where digital media technologies are more advanced and ubiquitous than they were when the latter was writing his book, in other respects their ideas are not too dissimilar. Bauman's contrasting of 'light modernity' with 'heavy modernity' is similar to Harvey's theorization. Light modernity is based on 'knowledge work', where employers are only committed to employees over the duration of their, often very short, contracts (Bauman 2000: 116–117). This is made possible by the conquering of time over space:

People who move and act faster, who come nearest to the momentariness of movement, are now the people who rule. And it is the people who cannot move as quickly, and more conspicuously yet the category of people who cannot at will leave their place at all, who are ruled. Domination consists in one's own capacity to escape, to disengage, to 'be elsewhere', and the right to decide the speed with which all that is done – while simultaneously stripping the people on the dominated side of their ability to arrest or constrain their moves or slow them down. The contemporary battle of domination is waged between forces armed, respectively, with the weapons of acceleration and procrastination.

(Bauman 2000: 119–120)

This last contribution comes close to Castells' concept of 'timeless time', which here will be deployed as a means of explaining the 'financialization' of the global economy.

The financialization of the world economy

Castells' (2010a) network society comprises global flows of information which travel through the nodes of the network. These nodes are sophisticated cities or regions which have systems of transportation and telecommunications advanced enough to facilitate these global flows. These flows take many forms, for instance they can be financial, knowledge-based or even criminal. But different flows will produce different networks: a global financial network will have New York, London, Frankfurt, Hong Kong and Tokyo as its main nodes; a global illegal drugs network might comprise of nodes in Alto Huallanga in Peru, Bogota, Miami and Amsterdam (Castells 2010a: 444–445). What Castells refers to as the 'space of flows' dominates the 'space of places' located outside the network. To give an example, the UK's second-largest city, Birmingham, is, in relation to the international finance industry, a space of place largely outside the global financial (space of) flows and hence in this sense at least, despite being around 100 miles away and in the same country, does not have as much in common with London as London does with Tokyo.

Castells posits two different temporal regimes for the space of flows and the space of places. In his formulation, 'timeless time' reigns in the space of flows, while our more familiar clock-time governs activity in the space of places. In other words, in global networks where decisions are made simultaneously and where time zones have been transcended, it seems as if time has become compressed to the point where it has detached itself from the clock-time to which most of us who live in the space of places are subject. It is not surprising that Castells (2010a: xli) first identified this phenomenon when he was doing research on the financial networks:

Speed became essential in financial transactions. The more capitalism went global, the more differences in time zones made possible the proliferation of interdependent financial markets to ensure the movement of capital around the clock. And so, a new form of time emerged in the financial markets, characterized by the compression of time to fractions of a second in financial transactions by using powerful computers and advanced telecommunication networks. Furthermore, the future was colonized, packaged, and sold as bets on future valuation, and as options between various future scenarios. Time as sequence was replaced by different trajectories of imagined time that were assigned market values. There was a relentless trend towards the annihilation of time as an orderly sequence, either by compression to the limit or by the blurring of the sequence between different shapes of future events.

(Castells 2010a: xl)

This trend was accelerated not only by the deregulation of telecommunication systems mentioned above but by the deregulation of financial markets around the world, beginning with the formation of the Eurodollars market in the mid-1960s and continuing with the abolition of restrictions on international currency exchanges in the world's great financial centres, including London, whose 'Big Bang' occurred in 1986 (Harvey 1990: 161, 163). As has been demonstrated above, the importance of advanced telecommunication systems was and is of such importance to financial networks that a great wave of capital was invested in the former after the financial deregulation of the 1980s (Lister *et al.* 2003: 197–198). Thus, global communications and networked economics are symbiotically linked, with the flow of capital to the telecommunications industry being reciprocated in the form of vociferous support for the present global financial architecture from the media outlets owned by those actors that gain so much from the system.

This powerful alliance (to which Castells (2010a) adds the world's political elite) perhaps explains why there is more than a hint in Castells's work of acquiescence in the face of a seemingly implacable global economic order. This might also be partly a consequence of his acceptance of the 'post-industrialism' of Daniel Bell and Alain Touraine. The emphasis in the earlier summary of theories of the post-industrial society on the importance of workers' developing their own theoretical knowledge and thus making themselves more self-sufficient is particularly pertinent to the digital economy. In this new type of economy, people are positively encouraged to be their own financial experts through activities like the buying and selling of shares. Digital media makes it easier for individual investors to manage their financial portfolios without the mediation of financial institutions. There is even software that makes this easier for traders (who can even make trades on their mobile phones), with the algorithmic HFT (high frequency trading) being particularly popular at the moment (Berry 2011: 156–159). While this is undoubtedly an attractive way of making money for those who are successful enough at this type of trading, it also carries considerable risks, which will be further explored in Chapter 5.

The transformation of existing industries and the growth of the creative industries

The financialization of the world economy is not the only prominent feature of the network economy. The network has transformed existing industries, while also driving the emergence of new ones, particularly those that are more knowledge-based or what we might term 'creative'. The transformation of existing industries can be linked directly to the re-organization of business operations elaborated earlier by Harvey. The modern corporation can now locate its operations in different parts of the network. This has led to a new organizational form, what Castells (2010a: 162–215) refers to as 'the network enterprise'. This enterprise is situated in a much more complicated global economic architecture than that which existed only a few decades before:

> [it seems that] the new economy is organized around global networks of capital, management, and information, whose access to technological know-how is at the roots of productivity and competitiveness. Business firms and, increasingly, organizations and institutions are organized in networks of variable geometry whose intertwining supersedes the traditional distinction between corporations and small business, cutting across sectors, and spreading along different geographical clusters of economic units.
>
> (Castells 2010a: 502)

Castells (2010a: 502) argues that while this is a powerful global economy powered by transnational capital, the large multinational corporations no longer reign supreme. Rather, it is the businesses or even individuals who can negotiate their way through this complex of interconnected networks that will prosper.

If we take this at face value, then we can identify positive elements of this new economy. In this more meritocratic environment, increased competition can be said to shake up markets dominated by complacent and inefficient monoliths. The success of relatively young companies like Amazon and eBay in shaking up bookselling and general retailing demonstrate the opportunities available for those with the chutzpah to grab them. There is benefit for the consumer too, as prices are driven down and companies can respond rapidly to changes in taste. Wal-Mart is probably the most successful example of a company which has used global networks to build a hugely sophisticated supply-chain that is highly responsive to signals sent from any of its points. Using RFID (radio frequency identification microchips) tags enables Wal-Mart not only to communicate internally, but with its suppliers too. So, for instance, when a cashier in a Wal-Mart store scans a purchased product, a signal is sent to the supplier, even if that supplier is based in another country. The supplier will then know that that particular item needs to be replaced and will begin that process immediately (Friedman 2006: 151). Wal-Mart's supply-chain works better than other companies because it is brilliant at coordinating various global networks. The information generated by the RFID chips can also quickly identify changes in consumer tastes or enable the company to reconfigure the supply-chain for particular events like, for instance, an incoming hurricane (Friedman 2006: 161–162). Of course, one should not deny that this is good for Wal-Mart's profits. Nonetheless, there are benefits for consumers too, as there are fewer instances of stores running out of stock of a certain item and consumers also get the feeling that their frequently changing whims are being catered for.

Spanish fashion retailer Zara's business is also heavily dependent on global information networks. In particular, it uses networked personal digital assistants (PDAs) to enable store managers to send information about customer preferences to a central planning office. Renowned for the frequency with which it introduces new designs into its stores, it takes Zara no more than 30 days to process products from the design stage to the shelf (Friedman 2006: 154). The example of Zara leads to a consideration of how the digital economy has created new industries and reinvigorated older ones. In particular, the digital economy has been linked with the so-called creative industries, of which fashion and design are part.

The 'creative industries' as a collective term was first adopted by the UK's New Labour government (1997–2010) and was a means by which it signalled its policy priorities in a post-industrial world. This policy was manifest in its embracing of the 13 industries it designated as creative (advertising; architecture; art and antiques market; crafts; design; designer fashion; film and video; interactive leisure software; music; performing arts; publishing; software and computer services; television and radio (Department for Culture, Media and Sport [UK] 2001)). As we saw earlier, Harvey (1990) had identified the importance of these creative industries – like advertising, computer software and design – in the so-called new economy, but did so from a critical perspective. Despite its history in the manufacturing labour movement, New Labour unashamedly embraced this postmodern economy. This partly reflects the way in which, as even the most critical accounts cited earlier attest to, this new type of digitally oriented economy is viewed as teleological – Margaret Thatcher's famous pronouncement: 'there is no alternative' (to the neo-liberal economy) is not dissimilar to Castells's insistence that it is foolhardy to try to resist the logic of the network society. Given that New Labour was unwilling in this sense to turn the clock back by taking the kind of measures needed to revivify the UK's manufacturing base, then the creative industries and, *sotto voce*, the finance industry were promoted as saviours of the British economy. The UK's attempt to 'map' these industries – essentially identify them and measure economic activity within them as part of the process of creating policies to help their development – was followed by many other nations, a manifestation perhaps of the belief by advanced capitalist nations that their societies were similarly post-industrial (Flew 2012: 11).

Notwithstanding the difficulty of capturing them precisely (see White 2009), figures for the creative industries in the UK suggested that this new approach was largely successful. According to government estimates, during the early phase of New Labour's administration, the creative industries represented around 5 per cent of the UK's total national income, generated some £60 billion per annum in economic value added and employed approximately 1.4 million people (DCMS 1998; cited in Flew 2012: 9). Since these figures are from 1998, the year after it came to power, it could be argued that New Labour's embrace of the creative industries was simply an acknowledgement of the prevailing reality. But subsequent figures showed steady growth in these industries:

> Two million people are employed in creative jobs and the sector contributes £60 billion a year – 7.3 per cent – to the British economy. Over the past decade, the creative sector has grown at twice the rate of the economy [emphasis in the original].
>
> (DCMS 2008: 6)

And it was not only the country that first institutionalized creative industries policy that recognized the potential in this sector. The United Nations Commission for Trade, Aid and Development (UNCTAD) publishes comprehensive reports on the worldwide creative economy, one of which in 2010 observed that:

> While the traditional manufacturing industries were seriously hit [by the global financial crisis], the more knowledge-based creative sectors were more resilient to external shocks. In 2008, despite the 12 per cent decline in global trade, world trade of creative goods and services continued its expansion, reaching $592 billion and reflecting an annual growth rate of 14 per cent during the period 2002–2008. This reconfirms that the creative industries have been one of the most dynamic sectors of the world economy throughout this decade.
>
> (UNCTAD 2010: xx)

This is a significant development in that, in Harvey's and, to a lesser extent, Castells's accounts of these industries, it is the developed world that benefits while the developing world's main contribution is in the ready supply of cheap labour. These figures suggest that the benefits of the creative industries are spreading beyond the most economically advanced nations. For this reason, similar to the UK New Labour government's use of mapping documents to give credibility to these industries, UNCTAD's measuring of the creative economy is an integral part of a strategy to aggressively promote the creative industries. This is manifest in its drawing up of the *Creative Industries Country Study Series* as a part of a pilot project to strengthen the creative industries in five countries in Africa, the Caribbean and the Pacific region (UNCTAD and the Government of the Republic of Zambia 2011: iii). UNCTAD's motivation for doing so is not only economic, but also because it views these industries as representing 'inclusive' and 'sustainable development' (UNCTAD 2012: 1).

Not surprisingly, the driving force behind much of this creative work is digital media, with worldwide growth in this sector growing faster than any other (UNCTAD 2010: 158). A quick glance at the *Fortune 500* list of world's most profitable companies in 2013 shows Apple in the number two position, followed by a number of corporations whose business is in computerization in the top 20: Wal-Mart (included in my list because of its sophisticated use of global information networks), Microsoft, IBM, Intel, Google and Oracle (CNN Money 2013). As reported elsewhere by the author, if we look at 2007 figures for the UK alone, the software, computer games and electronic publishing sector was the largest component of its 13 creative industries, constituting 31 per cent of its exports and 640,900 people of the 1,978,200 in creative employment; of the 409,500 new jobs created from 1997 to 2007 in these 13 industries, over 260,000 were in software, computer games and electronic publishing (DCMS 2007: 1–2, 5–8; cited in White 2009: 339–340). A recent report argued that SIC- (standard industrial classification-) based methodologies employed by the UK government tend to under-estimate the size of the digital economy. Employing a new methodology that, it argued, better reflected the nature of these new industries, the report identified nearly 270,000 digital companies in the UK, comprising 14.4 per cent of the total number of companies throughout the country. Overall, the digital economy is responsible for 11 per cent of all jobs in the UK (Growth Intelligence/NIESR 2013: 5). Though he does not mention the global economic crisis, Hal Varian's (chief economist of Google UK, the corporation that funded the research) claim in the foreword of the Growth Intelligence/NIESR report that 'those firms that are part of the digital economy are, on average, growing faster than those firms that are not' (Growth Intelligence/NIESR 2013: 4) is similar to UNCTAD's argument that in times of economic uncertainty this is the sector that we should look to for growth. Furthermore, UNCTAD's report argues that digital media's intervention in the creative industries is manifest not only in the making of products but also in providing an infrastructure for the marketing and distribution of creative goods (UNCTAD 2010: 158). If the networked and informational character of modern economies is also added, then all this constitutes an ensemble of digital media whose economic influence is much more significant than it at first appears.

Four aspects of the new economy

These trends have led some to champion a new economy based on claims that it has the following essential unique features:

1. *Open source* – we are happy to share our ideas.
2. *Connectivity* – we can collaborate with many people in different parts of the network.
3. Massive amounts of information and data make *the economy more transparent.*
4. *Barriers to entry lowered* – anyone can start a business.

Taking each of these claims in turn, there is some evidence for the existence of 'open source' ways of working in the development of computer software. The code that makes up the world's most popular webserver, Apache, is freely available and free to use. This means that any programmer can improve the existing code and then post their amended version online for all to share (Shirky 2010: 114–115). This type of evolutionary development of software enables flaws to be quickly identified and corrected. The popular operating system Linux and the web browser Firefox are also successful open source platforms. Indeed, it could be argued that the World Wide Web itself is the example *par excellence* of an open source approach to development, with its inventor Tim Berners-Lee fighting successfully for a free and non-proprietary platform (Friedman 2006: 61).

Related to this is the means by which networked communications facilitate this open source method of working by enabling us to send and receive information almost instantaneously to massive numbers of people. The type of connectivity that has driven the development of open source software has also been used in other projects. An example of this is *Wikipedia*, an online encyclopaedia which, like open source software, uses the online crowd to rectify existing content flaws and add new material in a never-ending process. If this method can be applied to software and educational content, could it not be deployed in the wider economy? Some, like Tapscott and Williams (2007) indeed argue that this approach should be adopted by those seeking to survive in the digital economy.

If openness and connectivity is the default position of the new economy then that will mean that there is more information transparency. This is significant because of the importance accorded to information in the economic theories of figures ranging from Fredriech Hayek to Joseph Stiglitz. Speaking from a neo-liberal perspective, Hayek argues that information is central to the running of the free market economy but since it is impossible for any one individual, institution or government to have all the relevant information available, then it is best to leave economic decision-making to the free market. Former World Bank economist Joseph Stiglitz (2002: xi, 73–74) argued similarly that information was central to a smoothly functioning economy but that the free market was just as much bedevilled by 'asymmetries of information' as individuals and institutions. Though it is not often made explicit by them, proponents of the digital-media-empowered so-called 'weightless economy' base their theories on the capacity of modern information networks to overcome any assymmetries of information and to make sure that consumers and businesses alike benefit from 'perfect information' (Gates 1995: 157–183). However, as Stiglitz (2002: 73) persuasively argues, it does not seem possible ever to eradicate all asymmetries of information.

Which businesses or, perhaps more pertinently, individuals will benefit in this open sourced, connected and transparent new economy? The answer, according to Tapscott and Williams (2007: 15), is potentially anyone:

> For individuals and small producers, this may be the birth of a new era, perhaps even a golden one, on par with the Italian renaissance or the rise of Athenian democracy. Mass collaboration across borders, disciplines and cultures is at once economical and enjoyable. We can peer produce an operating system, an encyclopedia, the media, a mutual fund, and even physical things like a motorcycle. We are becoming an economy unto ourselves – a vast global network of specialized producers that swap and exchange services for entertainment, sustenance, and learning. A new economic democracy is emerging in which we all have a lead role.

There is some empirical evidence to support this contention, in the form of the increasing success of self-published authors and many contemporary examples of music bands that have used the World Wide Web rather than traditional channels of publicity and distribution in their rise to prominence (Anderson 2007: 75–79, 99–106). Anderson (2007) has tried to capture this new dispensation in his concept of the 'long tail'. In summary, the long tail describes an economic environment in which production, distribution and marketing costs have been reduced so much by digital media that virtually anyone can produce cultural artefacts for the mass marketplace. New search technologies enable consumers to identify a much wider range of material too. In the media and cultural marketplace this, then, has led to the rise of the niche market and the diminution in power of the blockbuster produced by mass corporations. Pareto/Zipf's rule that 20 per cent of products account for 80 per cent of revenues (the 80/20 rule), seems to have been subverted by online markets, where, for instance, in a study of women's clothing by MIT, the bottom 80 per cent of products represented 28.8 per cent of sales (the offline (catalogue) figure was 15.7 per cent) (Anderson 2007: 135–136). Similar to the impact of the personalized financial trading software packages and websites which were mentioned earlier, the long tail appears to be enabling individuals and small companies to prosper in the creative industries.

Conclusion

As signposted at the beginning, this chapter has presented the most positive arguments for the existence and benefits of the digital economy. The next chapter will subject these claims to some critique. But, before that, I will introduce one final example which, while ostensibly demonstrating the strength of the digital economy, will also hint at one particular inherent weakness. In the process of writing his book on the global economy, Thomas Friedman (2006: 16) received an email from the then president of Johns Hopkins University, Bill Brody, which explained that many smaller and medium-sized hospitals in the USA are sending CAT scan images for analysis to radiologists in Australia and India. This story tells us something about network technologies, as the standardized protocol of the digital images means that they can be sent anywhere and can exploit the differences in world time to effectively outsource the work during the middle of the night, USA time (Friedman 2006: 16). This makes the whole process more efficient and undoubtedly cuts costs for the hospitals, especially when they are working with Indian radiologists whose wages are much lower than their USA counterparts. However, what happens to radiologists in the USA? Most likely, some will either lose their jobs or be pressured into working for lower pay and conditions. This anecdote highlights Castells's argument that work in the network society has been transformed in three distinct ways. Alongside more knowledge workers and a new global division of labour is the telling observation that wages in the USA definitely and the developed/industrialized world probably have stagnated, especially among the middle classes (Castells 2010a: xxi). The reason why this is will form part of the next chapter, a critique of the financialization of the world economy.

5
The Digital Economy and the Global Financial Crisis

Introduction

While the previous chapter contained some critique, mainly through using David Harvey's work on the postmodern economy, its overall intention was to present the strongest possible arguments for the efficacy of the digital economy. In the interests of balance, this chapter presents a much more thorough and sustained critical inquiry into the workings of the digital economy. The final paragraph of the previous chapter brings to mind the following exchange, which took place after a machine to automate tasks in a car assembly plant was introduced, nearly 60 years ago:

> In 1955, during a much publicized appearance at the General Motors Cleveland engine plant where the machine had recently been installed, GM CEO Charles E. Wilson, envisioning the fully automatic plant, is said to have asked United Auto Workers president Walter Reuther, 'Who's going to pay union dues?' – to which Reuther shot back, 'Who's going to buy your cars?'.
>
> (quoted in Aronowitz 2001/1994: 138)

This exemplifies the essence of the Fordist economy, as outlined in the previous chapter, namely that the workers needed a decent wage in order to properly fulfil their role as consumers of the manufacturing products disgorging from the very factories in which they worked. In this sense, excessive automation would upset the carefully constructed worker/consumer persona on which Fordism was dependent.

Given Castells's claim at the end of the previous chapter that wages in the most industrialized nations are stagnating, then the first question this chapter will ask is: what is the status of the average worker in the digital economy? The response to this question will involve an examination, not only of wages, but also of the concern expressed by some that automation and the move towards an economy based on symbolic rather than manufactured goods means that jobs will be lost and not easily replaced in other parts of the economy (Ford 2009). This will lead to a more general question about who makes the most money in the digital economy. One group of people who have certainly benefited are international financiers. To what extent, though, has the global financial crisis undermined their financial power? Another way of asking this question is: is the global financial crisis merely a periodic blip or does it constitute a more fundamental and inherent problem with the digital economy which cannot be fully resolved within the existing system? The section on the role of digital media in triggering the financial crisis will attempt to address these issues. If the digital economy is found to have inherent flaws, then how might we think of an alternative to this form of economic development? Here, it will be argued not only that there is an alternative to a totalizing digital economy, but that it already exists in the continued importance of manufacturing to the world economy. This will lead to a critique of the four claims of the most ardent proponents of the digital economy identified at the end of the last chapter.

Work in the digital economy

The anecdote above illustrates the long-held fear, going right back to the nineteenth-century Luddites, that advances in workplace technologies will result in the replacing of workers by machines. While trying to ascertain what will happen to future employment patterns involves, for this author, far too much speculation, looking at past and existing trends can nonetheless tell us something about the link between automation and employment. Various authors have expressed their concern that the informational character of the present economy threatens increased joblessness over the long-term (Aronowitz 2001/1994; Brynjolfsson and McAfee 2011; Ford 2009; Zuboff 2001/1988). Arguing mainly from a Marxist perspective, Aronowitz (2001/1994: 142) asserts that the purpose of digital technologies in the workplace is to eradicate paid work. Brynjolfsson and McAfee's (2011) fears about job losses are tempered by their conviction that, in the USA at least, the same technologies will provide alternative forms of employment. While Zuboff (2001/1988) is unsure where these technological developments will lead, she, like Ford (2009), is concerned that the character of these technologies is so qualitatively different from previous iterations that we cannot assume that, as in the past, they will create as many jobs as they destroy. While she identifies a specific aspect of these digital technologies – that they *informate* (collect information) as well as *automate* (Zuboff 2001/1988: 130) – her overall argument bears some similarities to Ford's (2009: 100–103) recycling of the old saw about exponential technological growth leading to the singularity. Before discussing these arguments in greater depth, it would be useful first to turn to some empirical evidence of past employment trends relating to the incorporation of disruptive technologies in the economy.

In a lengthy discussion on the subject, Castells (2010a: 272–275) reviews numerous studies of the relationship between workplace technologies and job losses and finds them generally unable to establish a causal link. The biggest survey, carried out by the Organisation for Economic Co-operation and Development (OECD) in 1994, found that:

> Detailed information, mainly from the manufacturing sector, provides evidence that technology is creating jobs. Since 1970 employment in high technology manufacturing has expanded, in sharp contrast to stagnation of medium and low technology sectors and job losses in low-skill manufacturing – at around 1% per year. Countries that have adapted best to new technologies and have shifted production and exports to rapidly growing high tech markets have tended to create more jobs … Japan realized a 4% increase in manufacturing employment in the 1970s and 1980s compared with a 1.5% increase in the US. Over the same period the European Community, where exports were increasingly specialized in relatively low-wage, low-tech industries, experienced a 20% drop in manufacturing employment.
>
> (OECD 1994: 32; cited in Castells 2010a: 279–280)

In addition to this, Castells (2010a: 268) relates that it was Japan and the USA, arguably the two most technologically advanced nations at that time, which created the most new jobs in the 1990s.

The old model of jobs that are lost through automation being replaced elsewhere in the economy can be illustrated with the example of agriculture. The main reason why agricultural employment has fallen so drastically in the world's most industrialized nations is because of the replacing of much manual labour with technology. Thus, in the UK agricultural employment fell from 50 per cent of the overall workforce in 1780 to a mere 2.2 per cent in 1988; during the same period productivity per capita increased 68-fold (Castells 2010a: 267). Those 'lost' workers migrated into other industries, first manufacturing and, latterly, the service industries. These figures, though, do not in and of themselves refute the arguments of those who argue that exponential growth in digital technologies is qualitatively different from technological change in the past (Ford 2009). We have already seen in the previous chapter how sending digital images of CAT scans overseas could cause potential job losses among radiologists in the USA and other industrialized nations. Ford (2009: 63–67) uses this example to illustrate that automation has become so sophisticated that, unlike in times past where workers performing routine tasks were the ones who were replaced, now it is the most highly valued and skilled jobs which are primarily threatened. Indeed, in comparing the work of a radiologist with a house keeper, Ford shows that it is the latter's job which is much harder to replicate than the former. Generally – and what might appear to some as counter-intuitively – Ford (2009: 67–73) argues that it is knowledge workers who are more vulnerable to automation than people whose jobs are less complex. Brynjolfsson and McAfee (2011) agree with Ford's diagnosis but argue that jobs can be created through innovation and entrepreneurship. The detail of how this might be done, though, is lacking; Brynjolfsson and McAfee (2011: 65–70) argue that their own country, the USA, should boost the education and skill levels of its workforce, increase investment and cut 'red tape' and taxation, but do not provide much in the way of empirical evidence to support the contention that these measures would lead to a growth in jobs. However, if Ford is correct about the future trends in automation, this would merely create unrealizable expectations that could not be fulfilled. Furthermore, as Chang (2011: 178–189) has demonstrated, beyond a certain optimal level there is little evidence that higher educational attainments and better skills among the working population cause additional productivity growth. While providing a verdict on Ford's proposition would be to indulge in a futurology that I think is suspect in academic argumentation, it does seem to have some credibility. Even Manuel Castells, an optimist in this debate, (2010a: 272), argues that there could be significant job losses throughout the economy if 'expansion in demand does not offset the increase in labour productivity'. Given that perpetual demand is likely eventually to lead to mounting environmental problems and that in any case the global financial crisis has somewhat depressed it, then Ford's scenario is not as implausible as it might seem.[8]

Brynjolfsson and McAfee's (2011: 56–58) argument that the digital economy must have a higher number of entrepreneurs and small businesses than the so-called old economy does describe a scenario that is already largely in place. Castells (2010a: 281–296) has detected a marked rise in the number of flexi-workers in the world economy along with the attendant job insecurity that this form of work brings in its wake. This has been exacerbated by the spread of mobile digital technologies, which make workers contactable at any hour of the week. The 'always on' worker has to cope not only with the increasing stress of the insecure workplace but also with the incursion of work into his/her home life (Miller 2011: 62). This is especially the case in the creative industries where, while entrepreneurship is perhaps encouraged in these industries more than any other, workers lead a somewhat 'precarious' existence (Neilson and Rossiter 2005).

Who makes money in the digital economy?

Focusing on the precarious character of working conditions in the digital economy generally and the creative industries specifically should not, though, be at the expense of an examination of the types of people and organizations that make enormous sums of money in this economy. In answer to the question: who makes money in the digital economy, some answer: everyone! This answer might be confined to the denizens of *Wired* magazine, but its writers – who have included Kevin Kelly, Chris Anderson, Nicholas Negroponte, the Electronic Freedom Frontier's John Perry Barlow and Internet pioneer Stewart Brand – are closer to the zeitgeist of American society, particularly its republican libertarian strain, than it appears superficially (Turner 2006: 222–232). A series of articles in the magazine's September 1999 issue was illustrative of this line of thinking. In his article, Kelly argues that there would be a massive increase in personal prosperity in the USA in the coming decades, such that most Americans would be very rich. And the reason why he believed this would come to pass was the strength of the digital economy:

> How many times in the history of mankind have we wired the planet to create a single marketplace? How often have entirely new channels of commerce been created by digital technology? When has money itself been transformed into thousands of instruments of investment? It may be that *at this particular moment* in our history, the convergence of a demographic peak, a new global marketplace, vast technological opportunities, and financial revolution will unleash two uninterrupted decades of growth [emphasis in the original].
>
> (Kelly 1999)

It would be presumptuous to claim that all these writers are wedded to the neo-liberal project: elsewhere, Kelly (2009) has praised the 'new socialism' that he believes digital culture is bringing into being. However, as Fuchs (2008: 301) has also recognized, their core argument that the free market can solve all economic ills and that the state should not intervene in the workings of the Internet places them, unwittingly or otherwise, firmly in the neo-liberal camp.

The reason I make this point is not to set up some kind of 'straw man' that I can easily demolish. Rather, it is to emphasize the politics behind some of the most influential ideas on the digital media economy at this time. Others who are optimistic about the digital economy as presently constituted, like Clay Shirky, Yochai Benkler and Charles Leadbetter would not describe themselves as cheerleaders for global capitalism (Freedman 2012: 77), but there is a danger that they provide intellectual cover for those who are trying to prevent us from delving too deeply into the question of who benefits from the digital economy. While *Wired*'s answer to this question is 'everyone', empirical evidence indicates that it is actually those who are successful in the traditional economy who are best placed to also prosper in the digital economy.

To convince us that the digital economy really is as different from the traditional economy as they say it is, its most zealous proponents claim that many elements of the classical economy are simply not relevant today. Chris Anderson's long tail thesis, for example, implies that monopolies will become less common as the Internet enables people to choose products from a wider range of sources. In actuality, monopolies are abundant in the digital economy. And many of those monopolies simply replicate those that exist offline. A report into online news from 2001 to 2006 found that it is dominated by the traditional news agencies, especially Associated Press (AP) and Reuters (Paterson 2006); there is no reason to suppose that this does not continue to be the case. This tendency towards monopoly also features in the relatively new businesses of Internet search, social networking and online advertising. Google's share of the search volume in New Zealand, the UK and Australia is 92, 90.5 and 88 per cent, respectively (Freedman 2012: 88). Figures from 2007 show that four companies – Google, MSN, Yahoo! and AOL – received 85 per cent of total revenue in online advertising in the USA (Freedman 2012: 88). And, contrary to the long tail thesis, this tendency towards concentration is evident in consumer choice too: 'Anita Elberse found that the top 10 per cent of songs on the digital download service Rhapsody accounted for nearly one-third of all plays: a result that demonstrates "a high level of concentration" ' (Elberse 2008: 2; cited in Freedman 2012: 90).

These figures seem unusual considering that digital media businesses supposedly have few barriers to entry and very low start-up costs. But is this an accurate characterization of business activity in this sector? Cultural products *are* characterized by low variable costs, in the sense that once the original is completed then it does not require much additional financial outlay to reproduce multiple copies. But the fixed costs, that is to say how much it costs to make the original, tend to be high (Hesmondhalgh 2007: 18–21). This is partly because there will always be many more failures in the cultural market than successes. Therefore, someone like a musician might initially spend a lot of time and money recording many songs which are not successful before finally making a breakthrough. Writing a book or making a film takes a very long time and can be financially costly, and yet there is no guarantee that the final product will be commercially successful. One way of offsetting this risk is to build a repertoire of products to ensure that, even if most of them fail, money will still be made from the handful that are successful. Large corporations are the best vehicles for hosting large repertoires and hence the reason why there is increasing concentration in the cultural/creative industries (Hesmondhalgh 2007: 22).[9]

While there might be increasing concentration in the digital media market, as shown in the examples above, some argue that these companies are different from the corporations that dominate the mass media environment. Chris Anderson (2009), for instance, emphasizes that search engines and social media websites are offering their content to users for free. For many writers like Jeff Jarvis, it is not a concern that Google is a monopoly because it operates much more ethically than other corporations. Utilizing Google's 'don't be evil' slogan in a worryingly uncritical manner, he argues that 'the costs of evil are starting to outweigh the benefits… When people can talk openly with, about, and around you, screwing them is no longer a valid business strategy' (Jarvis 2009: 102; cited in Freedman 2012: 77). But, as Freedman (2012: 85–87) notes, not surprisingly perhaps, Google is pretty much like any other large media corporation in its operation. The way in which Web 2.0 companies operate even brings to mind a term that was supposedly consigned to the dustbin when the mass media was eclipsed by digital media: gate-keeping. Mass media employed gate-keeping in three stages (news gathering, publishing and managing reader responses with strict editorial controls) to control content (Rettberg 2008: 103). With its plethora of news gatherers, ease of publication and open commentary platforms, there is almost universal consensus that the growth of digital media has resulted in the eclipsing of gate-keeping. However, we have already seen that a small number of the world's news agencies provide news content online – if you doubt this, just go to the news portal of one of the 'new' media organizations like Yahoo or Google and see how many of their news stories are sourced from these agencies. Similarly, if you want to download music legally online or buy a book then you are likely to do so through a 'gate-keeper' company like Apple or Amazon (Freedman 2012: 91).

But if the question of who makes money in the digital economy is a little clearer in relation to institutions, what about at the level of the individual? In other words, how many people are employed in these digital media corporations and how much are they paid? Jaron Lanier (2013: 49) argues that it is the 'Siren Server' – 'an elite computer, or coordinated collection of computers, on a network' – that gains most from the digital economy. In other words, companies which have been mentioned throughout this book like Facebook, Google, Amazon and Apple. As was noted above, some of these companies (in the example of the previous sentence, Facebook and Google) do not charge consumers for content, so under what economic model do they operate? The first important thing to note is that those companies that do not charge consumers also do not pay contributors for content. Instead, they make money from advertising. While it appears that consumers are taking advantage of all this free content, it could be argued that they are actually being exploited, in that they are allowing the content they create to be monetized by advertisers. For Lanier (2013: 10), this goes beyond economic exploitation to encompass invasion of personal privacy: 'cryptography for techies and massive spying on others'. This enables the kind of personalized advertising that was described in Chapter 2. While this type of service is often useful and only occasionally irritating, the principle of surrendering our privacy to siren servers is potentially more damaging. (The implications of this type of corporate surveillance will be discussed in Chapter 8.)

What about those siren servers, like Amazon and Apple, which *do* charge consumers for their products? These servers are deemed valuable in their cutting out of the so-called 'middle-man', the layer(s) of personnel that stand between you and a purchase. The metamorphosis of the giant US record company *Tower* from a number of imposing buildings in various parts of the world to an online store and international franchise is indicative of what is happening in the music industry today (*Banderas News* 2006; Keen 2007: 97–105; Rap News Network 2006).

In many respects, this is great for consumers in that they pay much less per song and do not even have to travel downtown or even to another town to make their purchases. However, as discussed earlier, the consumer in the Fordist economy needed a good, well-paid job as well as lower prices. In this sense, what some champions of the digital economy refer to as the 'middle-man' is actually a whole layer of employees that cannot, unless they find alternative, well-paid jobs, properly fulfil their roles as consumers. 3,000 of these employees lost their jobs as a result of the liquidation of Tower Records' US record stores (Keen 2007: 101). And it was not only giants like Tower who were felled, some 800 independent record stores in the USA closed between 2003 and 2006 (Keen 2007: 100). Independent record labels are likely to be adversely affected by this trend, especially as it was estimated that Tower alone was responsible for around 40–50 per cent of their market (Keen 2007: 104). Paradoxically, all this benefits not only new gate-keepers like Apple but also the old ones, in the form of the major record labels which, like many creative companies, consolidate as a means of managing risk and uncertainty in the market (Freedman 2012: 91; Hesmondhalgh 2007: 22; Keen 2007: 105).

To indicate what all this means for patterns of employment and wages in the digital economy, the following table of traditional manufacturing, networked retail and digital media companies has been constructed (Table 5.1):

Table 5.1 Statistics for employees, revenue and profits of Global 500 corporations in the digital economy, 2011

Corporation	Employees	Revenue (millions $)	Revenue/employee
Wal-Mart	2,100,000	421,849	$200,880
General Motors	202,000	135,592	$671,248
Ford Motor Co.	164,000	128,954	$786,305
Apple	49,400	65,225	$1,320,344
Amazon.com	33,700	34,204	$1,014,955
Google	24,400	29,321	$1,201,680

Source: CNN Money (2011).

The most noticeable aspect of this table is how much the networked retail corporations (Wal-Mart, Amazon and Apple) and the one digital media company (Google) diverge from the two traditional manufacturing giants, Ford and General Motors. The latter two employ an impressive number of employees and, judging by the revenue/employee ratio are able to pay their workers a decent median wage. Wal-Mart – which remember is the exemplar of a networked organization – employs a huge number of people, but revenue per employee is only $200,880. Once capital, taxation, profits, administrative and distribution costs are taken out of that figure, one wonders how much money is left for each employee; *Bloomberg* claims that wages at Wal-Mart are so low that state government programmes provide around $5,815 of subsidies per worker each year (this is hardly a ringing endorsement of the digital economy and the so-called transparent, self-sufficient free market!) (Ritholtz 2013). At the other extreme, Apple, Amazon and Google each employ less than 50,000 people but generate in excess of $1,000,000 for each of them. Astonishingly, Apple's profits are more than double General Motors' and Ford Motor Company's, and almost as much as Wal-Mart's, but the company employs less than one-third of the leanest of these companies (CNN Money 2011). Generalizing from this, admittedly small, though significant sample, two trends can be identified: the networked organization like Wal-Mart can supply huge numbers of jobs, but at seemingly low rates of pay; digital media companies can generate huge amounts of revenue which seemingly enrich greatly the relatively small number of people that they employ. This validates Lanier's (2013) view that the digital economy benefits mainly a small number of people, whose rewards are completely disproportionate to wages in the wider economy. It also seems to confirm Castells' (2010a) view that the digital economy widens the gap between the richest and poorest members of society. As Ford (2009), Lanier (2013) and Castells (2010a) all argue, this trend affects the middle classes – especially as more of the so-called 'middle-man' jobs are destroyed – as much as it does the poorest members of society, and these figures appear to validate this collective view.

The digital economy and the financial crisis

In August 2012 a company named Knight Capital lost $440 million in 45 minutes. What perhaps is most surprising about this is that this was not directly the result of human error or mis-judgement. In fact, no humans at all were involved in the trading that resulted in these losses; instead, Knight Capital was using algorithmic-based software that automates trading (Harford 2012). A crash that closed the NASDAQ stock market in New York for three hours was the most significant of a series of disruptions that occurred in the same month a year later (Garside 2013b). The rationale for automated trading differs from the motivation for automation in manufacturing and retailing. In the latter two industries, automation cuts personnel costs and, it might be added cynically, increases the number of operations that are not susceptible to industrial action. But while it is true that there is no longer any need for the frenetic trading pits that characterized 1980s stock market activity in places like New York and London (Knight 2013), there is little evidence that the financial institutions are concerned about either poor industrial relations or paying their employees huge salaries. The rationale for automation is solely about creating marginal gains over other competitors in this ferociously competitive environment. In online trading, speed is increasingly the most important commodity for those that want to make money (Adler 2012). And even the tiniest marginal advantage can yield enormous profits, as chief investment officer of Tradeworx Mani Mahjouri illustrated in a talk about trading in SPY securities and S&P (Standard and Poor's) futures:

> The prices of SPY and the futures contract generally move together but not in lockstep; as a rule, the futures contract tends to lead SPY by a few milliseconds. The reason is immaterial; all that matters is that the rule holds often enough for an algo[rithm] to make money by using changes in S&P futures to predict the direction of SPY over the next few milliseconds. How much money depends on how quickly the algo[rithm] can get data from Chicago to New York. As an experiment, Mahjouri applied this algorithm to the complete record of trades for an entire day and found that under optimum conditions – by which he meant transmitting data and orders at the speed of light – it would have made approximately 64,000 trades at an average profit of $0.0001 per share. The model specified a size of one share per trade; as a rule, high frequency traders deal in relatively small volumes, a few hundred shares at a time, because large orders perturb the market. Strictly as an illustration, consider that on an average day about 150 million shares of SPY change hands. Tradeworx claims to account for around 4 per cent of SPY trading, so by extrapolation, that would be 6 million shares of SPY, times $0.0001, equals...hmm...$600. By itself, this is a slow way to get rich. But multiply that figure by the number of such algorithms running at any given time – in the 'high hundreds' – and it starts to get interesting.
>
> (Adler 2012)

This combination of sophisticated algorithms and massive computational power, labelled high frequency trading (HFT), came to prominence during the Dow Jones 'flash crash' on 6 May 2010 (Berry 2011: 159; HPC Wire 2010). Mathematical and computational models are not new in the financial world. The Black-Scholes formula, for instance, was created in 1973 to price stock options and refined in the 1980s to cover the new derivatives that sprang up in that decade (Ford 2009: 44). But it is the specific capacity of modern networked computation which has given formulae so much power that they are beyond the capacity of human beings to comprehend (Adler 2012; Berry 2011: 159). Given the importance of the computerization of finance and its capacity to cause disruption on a global scale, it is surprising that only a handful of academics have devoted much time to studying it (Athique 2013; Berry 2011; MacKenzie and Spears 2012). While many authors (Castells 2010a; Shaxson 2011) have highlighted the way in which this industry has, in accumulating ever larger amounts of money for a small number of the global elite, threatened to run out of control, very few have focused on the role of digital media in exacerbating this problem. Castells (2010a: xix–xx) does discuss this in the preface of the 2010 edition of the *The Rise of the Network Society* but this accounts for only one of the six factors that he believes led to the crisis. That this was not discussed in the earlier editions of the book and is not elaborated on in a jointly written introduction that Castells wrote on the financial crisis in 2012, suggests that he has not developed as sophisticated a model of networked finance as he has of the network society (Castells, Caraca and Cardoso 2012). This lack of discussion in the general literature is all the more unusual when one considers that, as outlined above, this activity has grown during the post-2008 period, contrapuntal to the tighter regulation on other banking practices associated with the financial crisis.

For some, the continuing influence of automated finance is to be welcomed. As the previous chapter noted, fantasies of 'perfecting capitalism' through the 'perfect market' have animated techno-enthusiasts as far back as Bill Gates's 1995 (157–183) book *The Road Ahead*. If we cast our minds back to Hayek's belief that the perfect market is based on ignorance, insofar as the perfect market cannot be understood by one person, institution or government, then financial instruments that defy human comprehension are not problematic for those of a like mind. This type of neo-liberalism posits therefore that the financial crisis was not the result of too *much* technology, but of too *little*. The latest iteration of the belief that technology can produce the perfect market is related to algorithmic trading. And, like algorithmic trading, this belief has strengthened during the global financial crisis. This belief relies on old-fashioned positivism in its assertion that the complex cocktail of independent variables, human and otherwise, that make up the market can somehow be measured and tracked precisely. If that is so, then we must find the best way of enabling us to do this. This, though, can only be carried out by sophisticated computerized algorithms. The logic of holding this belief is that the fact that these past models have been inadequate to this task should not in any way prevent us from trying to create the perfect algorithm in the future (Weatherall 2013).

These ideas, that the marriage of computerization and algorithms can deliver the perfect market, are potent but fundamentally flawed. The first flaw in these arguments was foregrounded in the comment above about positivism. Classical economics is positivistic in its incorporation of a myriad of independent variables into standard formulae such as 'supply and demand'. These formulae are based on the idea that people act 'rationally' within the marketplace. But how do we define rational behaviour in the marketplace? An obvious answer to this question is that people will always try to secure the lowest price for any particular good or service – after all, why would you pay more than you should? Those who believe that the World Wide Web promotes price transparency through revealing price differentials much quicker than we could do by walking around all the stores in our home town (and it would be impossible simultaneously to physically look at all the other stores in other towns or even different countries) argue that it rewards the rational shopper and therefore improves the efficiency of the market. However, we do not always act rationally in the marketplace. For example, those who support sporting teams will pay ridiculous amounts of money to watch their heroes play and for the memorabilia associated with them. Sometimes, we simply do not have time to seek the lowest price; on other occasions we simply do not care. There might be cultural reasons or considerations of personal safety (I do not want to shop in that part of town or use that dodgy-looking website) that might prevent us from securing the best price. The rise of behavioural economics is a reflection of this growing realization in mainstream economics that human beings do not always act rationally in the marketplace (Kahneman 2011). The fact that classical economists almost to a person failed to predict the biggest global downturn since the 1930s demonstrates the shortcomings of their models; paradoxically, they were unable to identify the flaws in some of what we now realize as 'irrational' market practices, like making financial decisions on the basis that there would be unending growth in property prices.

Even if we took at face value the positivism of some financial writers, can we say that we have a more transparent market now than we had in the past? It is true that platforms like eBay and, in China, Taobao have improved transparency in particular markets, especially in their establishing of reputation systems which penalize unethical trading (Shirky 2010: 177). However, other markets still seem to be as opaque as they ever were. This is particularly the case with finance. As Castells (2010a: xx) has argued in relation to the global financial crisis:

> new financial technologies made possible the invention of numerous exotic financial products, as derivatives, futures, options, and securitized insurance (such as credit default swaps) became increasingly complex and intertwined, ultimately virtualizing capital and *eliminating any semblance of transparency in the markets* so that accounting procedures became meaningless.…*financial markets only partially function according to the logic of supply and demand*, and are largely shaped by 'information turbulences',…the mortgage crisis that started in 2007 in the United States after the bursting of the real-estate bubble reverberated throughout the global financial system [my emphasis].

In other words, there is a paradox in techno-capitalists celebrating transparency in markets for goods and services, while simultaneously seeking perfect algorithms that, in their embodiment of the Hayekian 'hidden hand', are deliberately opaque to the individual.

Unless we think that creating algorithms that no human being can fully understand is a sensible way of running the financial economy, the increase in the influence of this one sector, such that by 2007 it accounted for 40 per cent of all profits made in the USA, is worrying (Castells 2010a: xxi). But even this figure underplays the extent to which the economy as a whole has become increasingly subject to the logic of the financial markets (Castells 2010a). This means that it is virtually impossible to place a *cordon sanitaire* around these practices, and they seep into other economic activities where they have the capacity to greatly distort hitherto stable markets. It is here that Castells's observation above, that financial markets do not conform to the laws of supply and demand and lack transparency, assumes great importance. This is exacerbated by the fact that up to two-thirds of cross-border trade worldwide takes place *within* multinational corporations (Shaxson 2011: 12). This facilitates what can be termed 'transfer pricing' or, more appropriately, 'transfer mispricing', where by 'artificially adjusting the price for the internal transfer, multinationals can shift *profits* into a low-tax haven and *costs* into high-tax countries where they can be deducted against tax' [emphasis in the original] (Shaxson 2011: 11–12). The extent to which prices can become detached from their estimated worth can be illustrated with a few examples:

> a kilogram of toilet paper from China has been sold for $4,121, a litre of apple juice has been sold out of Israel at $2,052; ballpoint pens have left Trinidad valued at $8,500 each.
>
> (Shaxson 2011: 12)

Therefore, even in the age of eBay and Taobao, we are a very long way from price transparency not just in the financial industry but in the retail sector too.

But, like champions of the financial algorithms, proponents of friction-free capitalism might argue that global price transparency is possible if we continue to develop digital technologies in the correct way. While for a number of reasons that I do not have the space to discuss here, I reject this view, I will nonetheless discuss the implications below of just such an economic arrangement. What is surprising about many of the arguments of those who champion a Hayekian kind of frictionless capitalism is that they argue that this will benefit everyone. This is odd in the sense that even those of us who argue for equality of opportunity do not believe that it will lead to equality of outcome. Anyone who follows sport will know that some men and women will pull away from the rest of the pack even where everyone is given the same opportunities. That is fine in sport, where those of us who are not so proficient can go off and do something else. However, it is not possible to opt out of the economy in the same way, which is why it is more important to have appropriate restrictions on the accumulation of wealth by individuals and corporations, and welfare programmes to support those who cannot compete as well as others in the free market. The sloganeering that everyone will be equal in a frictionless capitalist society conveniently masks the extreme wealth of its many of most fervent proponents, as well as their neo-liberal economic views, and is the reason why we should not only challenge them on empirical grounds, but on ideological grounds too (Golumbia 2013).

Where does all this leave us in relation to the ongoing global financial crisis? Since the commercialization of the World Wide Web in the mid-1990s, we have witnessed the Asian financial crisis in the late 1990s, the dotcom bust at the turn of the millennium and the global financial crisis from 2008 onwards. While the link between the rise of the World Wide

Web and these crises might be nothing more than coincidence, it is worth observing them in the context of the digital economy arguments that have been reviewed in this chapter. And we can identify features of these crises which do relate to some of the arguments put forward here. During the 1997–1998 Asian financial crisis, the economies of South Korea, Thailand and Indonesia were damaged by the type of currency speculation encouraged by deregulated and computerized global financial markets; countries like China, which were not plugged into the network to anything like the same extent, did not suffer as much (Jacques 2009: 157). The dotcom crisis of 1999–2000 saw the NASDAQ fall by 40 per cent in less than a year. This crisis was precipitated by the massive over-valuation of Internet-related stocks, no doubt partly as a result of hyperbole about the digital economy similar to that identified in this chapter; these over-valuations also validate the earlier point about the increasing disjuncture in the modern economy between price and value (Lister *et al.* 2003: 211–212). The latest financial crisis, the most serious of all these crises, was precipitated partially, if not mainly, by a computerized global financial system that had careered out of control (Athique 2013: 140–141; Castells 2010a: xix–xx). Indeed, what makes this crisis more potent and deep-rooted than the previous ones is the existence and continuation of so-called 'dark pools' of technological trading systems (which sometimes employ 'dark algorithms') that are hidden from the regulators (Berry 2011: 160):

> Dark pools are a private or alternative trading system that allows participants to transact without displaying quotes publicly. Orders are anonymously matched and not reported to any entity, even the regulators (Younglai and Spicer 2009). Thus, the mainstream exchange-traded market does not have any clue about the volume of transactions happening in this parallel market or the prices at which they are being executed. Obviously, price discovery on the mainstream market, without dark pools information, becomes inefficient. Moreover, transactions carried out in dark pools effectively become over-the-counter in nature as the prices are not reported and financial risks not effectively managed. More critically, these risks can spread like wild fire as we saw in collateralised debt obligations and credit default swaps markets.
>
> (Shunmugam 2010)

Worryingly, this kind of trading continued to rise during the crisis, constituting 7.2 per cent of the total volume of US exchanges from April to June 2009 (Shunmugam 2010). The general public is only really aware of the impact of these and other practices when some of the flash crashes mentioned earlier occur. That these have occurred relatively frequently since the beginning of the financial crisis suggests that the underlying problems caused by this type of computerized trading is a long way from being resolved.

We live in a post-industrial society – long live manufacturing!

These observations of the actual working of the global financial system not only undermine the arguments of the most zealous of the proponents of the thesis that the digital economy represents a progressive and decisive rupture with the past, they also challenge the idea that the leading capitalist nations are post-industrial. This is particular the case in the supposed post-industrial nation *par excellence*, the United States:

> In 1973, the US produced an estimated 22 per cent more manufactured goods per head of population than the UK. By 2000, the difference was 91 per cent. America is a great manufacturing nation. The same can no longer be said about this country [the UK].
> (Rowthorn 2001; cited in Elliott and Atkinson 2007: 82)

While Chapter 7 will reprise the arguments of those who believe that Barack Obama won the 2008 and 2012 US presidential elections because of his mastery of digital media, others believe that it was actually his bailout of General Motors and Chrysler which helped him win over crucial auto workers in Ohio and Michigan in the 2012 election; Indiana, where foreign brands like Subaru and Honda dominate, tellingly was won by his opponent, Mitt Romney (Kumar 2012; Woodward 2012). Figures from the 2013 Fortune 500 list of the USA's largest corporations emphasize the continuing importance of traditional manufacturing industries to the American economy (Table 5.2).

One of the most noticeable things in this table is the presence in the top ten of two motor companies, General Motors and Ford; if we live in a post-Fordist economy, then no one told Ford! The creative and computing industries that we associate with the digital economy are not faring so well. When one thinks of the success of the AMC TV series *Mad Men*, with its focus on New York's advertising industry, it is surprising that both the advertising and entertainment industries have such modest figures. While it might be argued that these types of creative

Table 5.2   Largest Fortune 500 companies in selected industrial categories, 2013

Industry	Company	Position within its industry	Position in Fortune 500
Motor vehicles & parts	General Motors	1	7
	Ford Motor Company	2	10
	Johnson Controls	3	67
Information technology services	IBM	1	20
	Xerox	2	131
	Computer Sciences	3	176
Entertainment	Walt Disney	1	66
	News Corporation	2	91
	Time Warner	3	105
Advertising & marketing	Omnicom Group	1	191
	Interpublic Group	2	366

Source: CNN Money (2013).

industries have been more susceptible to collateral damage from the global financial crisis than others, it is actually difficult to establish a causal link between the fluctuations of the finance industry and the state of the creative industries (Pratt and Hutton 2013). A study of the creative industries in New York during the downturn shows that manufacturing industries shed a greater percentage of their workforce than the creative industries (Indergaard 2013). Indeed, figures predating the financial crisis (from the 2007 Fortune 500 list) show that the advertising and marketing, and information technology services sectors have actually gained since then, while the entertainment sector has dropped back. As in 2013, General Motors and Ford were in the top ten, at third and seventh place, respectively, with Johnson Controls in the same (67th) position that it occupied in 2013 (CNN Money 2007). This demonstrates that, in good times and bad, manufacturing in the USA remains a robust sector, so much so that presidents, as Obama did in his bailout in 2009, are willing to provide financial support for its ailing industries.

Part of the reason for the continuing importance of manufacturing is that it has benefited as much from global networked communications as other industries (Castells 2010a). It is also difficult to disentangle manufacturing from the supposedly more creative industries. Take the example of design. This might be a creative practice, which has an intangible (in the sense that the final product is not a physical object) outcome. But for whom is the designer working? Ivor Tiefenbrun, who founded Glasgow-based sound systems company Linn Products, provides an answer:

> What I call design is so closely coupled to manufacturing and so competitive that it cannot survive if the links are too tendentious. The reason Britain has fewer in-house designers than competitor countries who have retained more of their manufacturing base is because our manufacturers are so debilitated that they cannot afford, or do not choose to invest in, sufficient in-house capacity.
>   (Tiefenbrun 2006; cited in Elliott and Atkinson 2007: 84)

But even though manufacturing industries in the UK have shed more jobs compared to most other industrialized nations (Castells 2010a: 224), most of the fall in manufacturing output can be put down to falling prices (PCs for instance cost a lot less than they did years ago), improvements in productivity relative to those in the service sector and the re-classification of industrial categories (Chang 2011: 92–94). Taking relative price changes into account, we can see that manufacturing total output as a proportion of all output showed only a gradual decline from 1950 to 1990, from 27 to 24 per cent (Alford 1997: 6). Furthermore, job losses in the manufacturing industries are mainly a result of outsourcing to other countries (Chang 2011: 92–93).

Conclusion of Part II: A critique of the four supposedly main aspects of the digital economy

Let us remind ourselves of the four supposedly main aspects of the digital economy that were identified in the previous chapter:

1. *Open source* – we are happy to share our ideas.
2. *Connectivity* – we can collaborate with many people in different parts of the network.
3. Massive amounts of information and data makes *the economy more transparent.*
4. *Barriers to entry lowered* – anyone can start a business.

The flame of open source ways of working still burns, especially in relation to the development of software as the examples of the success of programmes like Mozilla Firefox, Linux and Wikipedia show (Athique 2013; Berry 2011; Freedman 2012). But the logic of the move towards more informational economies generally and the commercialization of

Internet content specifically means that patenting and intellectual property legislation has colonized virtually all areas of the World Wide Web (Athique 2013; Freedman 2012). This has gone so far that it strongly challenges the notion that there are separate spheres of the World Wide Web, one commercial, the other non-commercial (Berry 2011: 61; Freedman 2012: 83). We can see how this plays out in relation to a company, Google, which is seen as more open than the more traditional form of corporation. Jeff Jarvis (2009: 4; cited in Freedman 2012: 73) was undoubtedly talking about Google when he said '[o]wning pipelines, people, products, or even intellectual property is no longer the key to success. Openness is'. However, Google is not going to reveal the algorithm that drives its search engine; indeed, why would it reveal the one thing that makes it the market leader?

Greater connectivity across the worldwide network has no doubt led to greater efficiencies. The billions of people from the former eastern *bloc* and China that have joined this network since the end of the Cold War have to a large extent benefited from the greater access to world markets that this has brought about (Castells 2010a; Friedman 2006). While the increasingly networked organization of the operations of multinational corporations has led to jobs being lost in the developed world, this has been offset by so many manufacturing jobs being created in rapidly industrializing countries such that there has been an overall *increase* in employment in this sector in recent decades (Castells 2010a: xxii). However, being typically non-unionized, many of these jobs pay poorly and exist within exploitative, if not downright dangerous, environments. Thus, this trend enables corporations to cut unionized jobs in developed countries like the United States and take advantage of lax labour laws overseas to cut costs. The story of Apple and the problem of the working conditions in the factories of its major supplier in China, Foxconn, show that even the seemingly most progressive companies work within this new model of globalized industrial relations (Apple Press Info 2012).

While online retail websites have, to a significant degree, given the consumer more price transparency, the claim that the digital economy as a whole is more transparent than the traditional economy cannot survive any serious scrutiny, as the work of Shaxson and Castells and others has demonstrated. Furthermore, the whole idea of seeking the 'perfect market' is a neo-liberal fantasy, which masks the great sums of money that its most zealous advocates stand to make at the expense of the vast majority of this planet's citizens if their dreams are actualized.

The fourth point about entry into the market being easier in the digital economy is valid only in isolated instances. While there are examples of small businesses and individuals making large sums from Internet-based commerce and a smattering of musicians and writers like, for instance the Arctic Monkeys and E.L. James, who have launched successful careers on the back of Internet content that has gone viral, it is hard to avoid the conclusion that these are the digital versions of a young footballer being 'discovered' by a scout who just happened to be walking past the match in which they were playing; in other words, rare phenomena from which no general rule can be applied (Keen 2007: 111–112; Sales 2013). Hesmondhalgh's earlier argument that the high failure rate of cultural products means that – if you are individual – you can spend an inordinate amount of time and money writing before you get a foothold in the market or – if you are corporation – spend a large amount of money developing a repertoire is as relevant to the digital economy as it is to the so-called conventional economy.

While this conclusion might seem unduly pessimistic, what it really emphasizes is that the concept of the digital economy does not deviate that much from our more conventional understanding of the economy. It is therefore incumbent on those who argue for new business practices in the digital economy to make suggestions that are more progressive than simply using intellectual property regimes to shovel large sums of money into the hands of ever fatter corporations. Virtual reality pioneer turned Internet critic Jaron Lanier has put forward the radical idea that every contribution to the sum content of the World Wide Web should receive a small payment. Though the practicalities of this would be fiendishly difficult to address, it would at least ensure that people were being paid for content rather than giving it away for free (Lanier 2013). It would be good if others with influence could similarly contribute their own ideas rather than leave the field free to turbo-charged neo-liberals with all their shortcomings.

Part III
Digital Media Use

Introduction to Part III

While the previous two parts conceptualized the politics and economics of digital media, I am aware of the shortcomings of relying solely on perspectives from these two fields. As someone whose PhD was in political science, it is not surprising that I think politics is an important field of study. While I hope I have convinced the reader that a political analysis is crucial to our overall understanding of digital media, I appreciate that many might not share my fascination with politics. Similarly, while I think that any study of digital media must get to grips with its basic political economy, there is always the danger that this type of approach veers into a Marxist terrain which subordinates our everyday use of this media (what Marxists call the superstructure) to the economy (what Marxists refer to as the base). This part, then, is written for those who want to look beyond politics and economics in their attempt to understand digital media. Given that I do not take a reductionist view of digital media as being merely the plaything of either politics or economics, I would like to explore how ordinary citizens actually *use* digital media. As we will see, as Part III unfolds, the way in which our use of digital media has evolved has in many respects defied even the designers of those technologies, and its proliferation beyond the most advanced industrialized nations highlights great differentials in practices across the globe.

In the chapter following this introduction, there will be a discussion of theories which attempt to conceptualize both the interaction of various forms of digital media with the social world and our use of them. This approach is taken to re-emphasize the point that technologies operate differently in different societies. This does not mean that our use of digital media is solely dictated by the society in which we live, but that we need to look beyond the intrinsic elements of those media if we are to construct a sophisticated theory of use. Having a grounding in these theories will provide a better understanding of the case studies that will populate chapters 7–9: the new social movements, surveillance and the use of digital media in the developing world.

6
Reading/Using Digital Media

Introduction

In the mass media age, theories of media use were dominated by two particular strands: instrumentalism and technological determinism. While very few studies of the way in which the audience used various media were *wholly* instrumentalist or technologically deterministic, most tended to be located somewhere on the continuum between these two positions. This chapter will critique these two positions before discussing theories that attempt to go beyond this crude binary, in particular Joost Van Loon's conceptualization of media use and Neil Postman's media ecology theory. While this chapter is mainly theoretical (with case studies following in chapters 7, 8 and 9), it will discuss how one of our oldest media practices, reading, has been affected by the proliferation of digital media. This particular case study will be theorized through Walter Ong and Marshall McLuhan's work on 'secondary orality'.

Instrumentalist and technologically deterministic approaches to media use

While, as stated in the introduction, instrumentalist and technologically deterministic approaches to media use are associated with the mass media, the persistence of these theories is evidenced by their continued use in the contemporary digital media environment. An example of technological determinism in this regard is evidenced in the tendency of commentators and politicians to ascribe causality to social networking technologies in relation to events like the English riots of August 2011 and the Arab Spring (see Keen 2012). So, while this chapter is dedicated to attempting to find better ways of conceptualizing our engagement with digital media than instrumentalism and technological determinism can provide, it would be useful to define those theories first before moving on to the main arguments.

As the examples above suggested, technological determinism is the belief that technology *determines* social outcomes or, to formulate it slightly differently, has *agency* or *causation*. The term tends to be deployed pejoratively and, even in the one instance that the author could find where it is embraced, it is done so with heavy qualification (Heilbroner 2003/1967). Despite this, it is clear that many analyses of both analogue and digital media could be considered technologically deterministic – of those theorists discussed in this book, Jacques Ellul, Manuel Castells and Marshall McLuhan have all on occasion been labelled in this way (Lister *et al.* 2003: 312; Miller 2011: 4). Whether or not we think that their views can be reduced to this simple ascription, that this term is used so often suggests that it has some importance beyond the merely pejorative. If we dig a little further, we will find that many accounts of the influence of all forms of media employ an element of determinism. Latter-day claims that mobile phones and social media are altering our behaviour are similar to claims in the 1950s and 1960s that television was turning people into 'couch potatoes' or, in cases where content was violent, inciting them to acts of aggression. For this reason, any analysis of digital media must at least engage with the concept of determinism, technological or otherwise, before recommending its supersession.

How can we explain the seeming contradiction between the commonly held view that technological determinism is fundamentally flawed, and the existence of many arguments from media and communication scholars which appear to be deterministic? Heilbroner's (2003/1967: 401) unashamedly technologically determinist stance is qualified by his invoking of William James's term 'soft determinism'; the philosopher's concept is also employed by the editors of a collection of papers from a 1994 symposium on technological determinism (Smith and Marx 1996; cited in Lister *et al.* 2003: 311). Soft determinism insofar as it relates to technology is the acceptance that ancillary factors, social in orientation, are also involved in determination or causality (Heilbroner 2003/1967: 401; Lister *et al.* 2003: 310–312; Wessels 2010: 31). Thus, it is clear that when people invoke technological determinism as a pejorative trope, they are referring to hard determinism, rather than the soft version. It must be stressed, though, that even with soft determinism, technology is seen as the main determinant of social change, with the other factors playing a secondary role.

In response to this, many have argued that it is society, rather than the intrinsic properties of a particular technology, which determines change. But in being a mirror image of technological determinism, it runs the risk of viewing technology as merely, in the words of Raymond Williams, 'symptomatic' of society (Williams 1990/1975; cited in Miller 2011: 4). A better expression of this mode of thought, especially when related to the study of the media is 'instrumentalism' (Van Loon 2008). Here is an early and very clear formulation of media instrumentalism:

> Instrumentalism may be defined as the view that social institutions are the instruments of those who occupy elite decision-making positions (corporate owners, state bureaucrats, senior politicians), and who manipulate them in their own narrow interests, often at the expense of subordinate social groups.
>
> (Hackett 1986: 141–142)

Instrumentalism not only relegates technology to a subordinate role, but citizens too are treated as dupes. As the quotation above intimates, readers, viewers and listeners are seen as reacting credulously to the worldview of powerful media owners. (A popular refrain goes something like: 'he/she believes everything he/she reads in the Murdoch press'.) In other words, technology is used as an 'instrument' by powerful interests to wield disproportionate influence over viewers, watchers and listeners. In the relegation of citizens to this subordinate role, whether it be at the hands of media technologies or the owners of those technologies, both technological determinism and instrumentalism seem to be flawed as conceptual tools for understanding media use.

Perhaps the biggest problem with these two schools of thought is not so much the over-emphasis on one factor as an explanation for a range of social effects, but the clinical separation of technology and society (Williams 1990/1975; cited in Miller 2011: 5). This separation is the reason why even soft determinism, whether of the technological determinist or instrumentalist variety, tends towards the excessive downplaying of the social and technology, respectively. This theoretical model, which places technology and society at opposite ends of a continuum, with each dominating the other at every point except in the exact middle, needs therefore to be replaced by one which focuses on the interaction between them.

How do we use digital media?

The complexity of the digital media environment can best be illustrated with the example of the mobile phone. The development of mobile telephony was long in gestation, up until the 1990s it was used in a similar way to fixed telephones, and was not particularly popular (Athique 2013: 115). But then, almost imperceptibly to the wider public, a change in use gradually came into being:

> The first signs of the next shift began to reveal themselves to me on a spring afternoon in the year 2000. That was when I began to notice people on the streets of Tokyo staring at their mobile phones instead of talking to them. The sight of this behavior, now commonplace in much of the world triggered a sensation I had experienced few times before – the instant recognition that a technology is going to change my life in ways I can scarcely imagine.
>
> (Rheingold 2002: xi)

Though he was commenting just over a decade ago, Rheingold's description of texting sounds like something from a different age. What his comment does reveal is the inability of even the most perceptive commentators on digital media to predict its future development. It is unlikely that anyone in the 1980s or 1990s could have envisaged how popular texting would become, let alone predict the myriad functions of today's smartphones. Explaining this development is difficult to do without employing some sort of technological determinism: 'After all, who felt the need to "text", to change a PC desktop's wallpaper or nurture a Tamagotchi virtual pet until a consumer device suggested we might?' (Lister *et al.* 2003: 232). This is especially so when we realize that none of these developments in mobile telephony would have taken place without the miniaturization of devices and the adoption of digital technologies which took place in the 1990s (Athique 2013: 115). However, the fact that it was not only digital media commentators but manufacturers too who failed to predict these precise uses suggests that there is an extent to which users are shaping practices too. (An example of the inability of manufacturers to keep up with modern trends can be seen in the eclipsing of the early market leaders Nokia and Blackberry by older technology companies like Apple and Samsung (Kuittinen 2013).)

The entertainment industry provides more examples, the first of which is three decades old. The eclipsing of Betamax video cassette recorders and tapes by VHS in the 1980s was not because of any intrinsic technical superiority (some have even argued that the Betamax was technically better), but because users decided that they preferred using that particular product (CIBER 2007: 11; The Work Foundation/Nesta 2007: 69). Patterns of usage in the contemporary film and television industry are similarly difficult to understand. There are now many different ways in which you can watch films and television programmes: on your PC, laptop, portable viewer and mobile phone, as well as on more conventional platforms. Logic dictates that users will gravitate towards the most technically sophisticated of those platforms. It is no surprise, therefore, that when I talk to my Chinese students about watching films or TV dramas, they are perplexed when I talk about DVDs, as the vast majority of them use websites like *Tudou*, *Youku* and *PPTV* to view content. Watching live sport online is also popular in China. The fact that China has a laxer intellectual property regime partly explains this pattern of use; as mentioned in Chapter 1, it is estimated that pirated content comprises up to 90 per cent of film and music watched and listened to in the country (Montgomery 2010: 108). In countries with much tougher intellectual property regimes and where proprietary software like iTunes dominates the market, watching and listening to pirated content is much less in evidence. It is still possible in the UK to watch online live football matches from the English Premiership, but tougher measures are being put in place to make this much more difficult for the viewer (Ziegler 2013). And even in China, where web piracy supposedly is rampant, there are huge DVD stores in its largest city, Shanghai. That people continue to watch content on this supposedly less sophisticated platform illustrates not only that they do not want to breach intellectual property laws, but also that the way in which we interact with digital media technologies often defies rational explanation.

In another branch of the entertainment industry, music, where piracy is probably even more common, there is also a variegated range of practices. A linear explanation of the development of platforms would go something like this: cumbersome and fragile vinyl records and tape cassettes were gradually replaced by CDs in the 1980s and 1990s, which in turn were replaced by digital files in the new millennium. While this is largely correct, this type of account unwittingly conflates innovation with quality – in other words, the newer technologies *must* provide better acoustic quality. But this is not necessarily so. Seidensticker (2006: 32) argues that the acoustic quality of CDs is better than that produced by MP3 players. Even vinyl records have defied authoritative pronouncements of their demise, being still used by DJs on mixing desks and even continuing to be sold in bulk in the mass market (Michaels 2013).

Is there a 'third way' between technological determinism and instrumentalism?

It would be easy at this stage to answer the titular question above with the rather banal: 'yes', developments in digital media are the result of a combination of technological progress and social trends. A more sophisticated answer would require analysing the work of two theorists who have provided the most convincing models to explain the interaction between media technologies and the social: Joost Van Loon and Neil Postman. Van Loon's general philosophy on this interaction can be summarized thus:

> First, the capacity to act, which media undoubtedly have, should not be confused by the capacity to *determine*. Second, even if media-technologies have the capacity to *determine*, this does not mean that they can do so unconditionally themselves, let alone on their own. In simpler terms, the fact that media are capable of acting, and perhaps even conditioning, does not mean that other forces (e.g., social, economic, cultural, political, physiological, psychological) are therefore irrelevant [my emphasis].
>
> (Van Loon 2008: 30)

In order, though, to fully comprehend the complexity of Van Loon's theory of media use, an understanding is needed of two key terms: *enframing* and *binding-use*.

Enframing is a concept which Van Loon takes from Heidegger's (2003/1977) use of *Gestell* in his famous essay *The Question Concerning Technology*. In Van Loon's view, *Gestell* 'is best translated as "ordering" in the double sense of providing a structure (putting into place) and commanding specific actions' (Van Loon 2008: 148). Van Loon emphasizes that enframing relates to whole technological assemblages, rather than individual technologies or applications. As in the quotation above, Van Loon does not believe that media technologies *determine* how we think, but their overall technological apparatus does order, enframe or set the parameters for the way in which we conceive the world around us (Van Loon 2008: 96). But this ordering can only be effective if is compatible with existing social practices.

In the routine practices of everyday life, we invoke technology as a set of tools. That is, we relate to these tools in terms of how they are supposed to work. It is exactly in such ordinary circumstances, when things that are ready-to-hand perform as expected, that technology is taken-for-granted and most effective in its enframing of the world. In contrast, when technology no longer functions as expected; when a danger (or risk) is being revealed, when we are confronted with discomfort and inconvenience, the *enframing* is no longer 'taken-for-granted' and technology no longer 'holds sway' [emphasis in the original].

(Van Loon 2008: 30)

This emphasis on technologies being ready-to-hand suggests that, for a technology to be successfully incorporated, it needs to be routinized, a concept that Van Loon (2008: 138) explains in terms of *binding-use*:

Technology-use becomes embodied through routine; routine is a habitualization of technology-use, by means of which it becomes binding. Our bodies are attuned to technology-use when we do things seemingly automatically, for example, when we apply the fine motor skills of holding a pen to write, when we brush our teeth, or use a computer keyboard.

But binding-use can only occur: 'if it attunes our *Dasein* to our being in the world. That is the attunement is not optional but necessary' (Van Loon 2008: 138).

The complex relationship between enframing, binding-use and the adoption of technologies can be simplified in the following way:

- **STEP 1**
- ENFRAMING – orders (structures and commands) a sphere of action (*Technology*) +
- **STEP 2**
- BINDING-USE – technology is only accepted once it becomes bound to a particular use (*User*)
- = SUCCESSFUL TECHNOLOGY

We can think of this in relation to some examples of successful and unsuccessful adoptions of technologies. The bicycle has evolved from very primitive and unwieldy forms to the sleeker 'racing' styles that we see today. The nineteenth-century penny farthing, with its huge front wheel, was in many respects not much different from bicycles today. In this sense it could be said have been successful in terms of enframing, in commanding 'a sphere of action', namely cycling. However, over time this particular configuration became 'dis-bound' from cycling. While it was initially popular, the penny farthing was replaced by bicycle designs which were much more amenable to the general public and hence became 'bound' to cycling (their 'use'). An example mentioned earlier, the mobile phone, can also be used to illustrate this point. The technology in the 1990s mobile phone enframed certain actions, such as emergency calls and text messages, but back in those days only the former could be said to constitute a use that was binding. However, the popularity of texting has grown since then to the point where it has more of a claim for a binding-use technology than does the emergency call.

There is, though, a third step that can be added to the enframing/binding-use model illustrated above, stemming from a consideration of the concept of media ecology. In nature, a particular species' significance cannot be considered without thinking about its position within the overall ecology. We might think about this in relation to the cane toad, which was introduced into Australia in the 1930s to eat the scarab beetles which damage sugar cane. The cane toad, however, did not eat only scarab beetles, but a whole range of other species too. Having few natural predators in Australia, cane toads have bred so rapidly that they have had a profound effect on the ecology of the nation, which is exacerbated by the poison that they emit to the preyed-upon and the predator alike (Australian Museum 2013). Neil Postman (1993: 18) argues that media technologies should be understood in similarly ecological terms.

> Technological change is neither additive nor subtractive. It is ecological. I mean 'ecological' in the same sense as the word is used by environmental scientists. One significant change generates total change. If you remove the caterpillars from a given habitat, you are not left with the same environment minus caterpillars: you have a new environment, and you have reconstituted the conditions of survival; the same is true if you add caterpillars to an environment that has had none. This is how the ecology of media works as well. A new technology does not add or subtract something. It changes everything.

While we might contest Postman's view that the advent of a new technology 'changes everything', if we accept that it at least alters the social environment in some significant way then it allows us to develop the enframing/binding-use model. The introduction of testing and statistics in education altered the way in which we think about students' intellectual ability to the extent that we can no longer comprehend how we could measure it outside these frameworks (Postman 1993: 123–143). Similarly, the introduction of increasingly sophisticated medical machines and drugs has left us almost completely dependent on them for diagnosing and treating all manner of ailments and diseases, even when they are ill-suited to do so (Postman 1993: 92–106). While the enframing/binding-use model developed up to this point emphasizes the importance of agency – in the sense that human beings must 'accept' a certain use of technology if it is to become successful – Postman's view is that the acceptance of these technologies subsequently circumscribes human agency. In other words, becoming 'bound' to a particular use should, as the word itself implies, constitute some sort of restriction. In relation to the cane toad, it could be said that human beings were responsible, in the form of physically releasing them into the Australian environment, for accepting it as a means by which to control pests. But the cane toad is 'enframed' in such a way that its interaction with the environment is largely beyond the control of human beings. An example of how this works in relation to media technologies is in our use of mobile phones. We are seemingly free to refrain from using them but is it really feasible to boycott this technology when all our friends and family communicate through mobile phones? Thus the model above can be altered with the addition of a third step:

- **STEP 1**
- ENFRAMING – orders (structures and commands) a sphere of action (*Technology*) +
- **STEP 2**
- BINDING-USE – technology is only accepted once it becomes bound to a particular use (*User*)
- = SUCCESSFUL TECHNOLOGY whose use is restricted by its enframing (**STEP 3**)

This model is particularly suitable for digital media technologies. Musician and inventor of Virtual Reality Jaron Lanier illustrates this with the example of the universal adoption of MIDI as a means of representing musical notes. MIDI's designer Dave Smith never intended it to be a universal standard, as he only wanted to expand the range of the sounds he was listening to by connecting a number of synthesizers together. However, a technology that cannot adequately capture the sound of a violin, singer or saxophone has become standard for all digital music; though not, it should be added, at the behest of its inventor (Lanier 2011: 7). The danger of what Lanier terms 'lock-in' is not a new phenomenon. Railroad gauges and uniformly dimensioned tunnels on the nineteenth-century London Underground left no room for air-conditioning to function properly, a problem that once 'locked in' continues to make commuting uncomfortable for contemporary workers in the UK's capital (Lanier 2011: 7–8). This is bad enough for older technologies.

> But software is worse than railroads, because it must always adhere with absolute perfection to a boundlessly particular, arbitrary, tangled, intractable messiness. The engineering requirements are so stringent and perverse that adapting to shifting standards can be an endless struggle. So while lock-in may be a gangster in the world of railroads, it is an absolute tyrant in the digital world.
>
> (Lanier 2011: 8)

There is a long line of examples of how digital technologies are configured in a counter-intuitive way, going all the way back to the nineteenth century and the QWERTY keyboard, which is still used on contemporary PCs, laptops and smartphones. Once these uses of technologies are locked in, further designs must conform to the standard, thus removing

> ideas that do not fit into the winning digital representation scheme, but it also reduces or narrows the ideas it immortalizes, by cutting away the unfathomable penumbra of meaning that distinguishes a word in natural language from a command in a computer program.
>
> (Lanier 2011: 10)

As Chapter 5 showed, in a world where proprietary software and standard operating systems are colonizing the World Wide Web and where monopolies are prevalent in key markets, such as social networking and search engines, pronouncements to the effect that the user is all-powerful should be viewed with a sceptical eye.

Reading digital media – the concept of secondary orality

One activity which has supposedly been altered radically by the proliferation of digital media and in which the user is purported to have greater autonomy is reading. As the chapter will go on to elaborate, some see this as a liberating experience, while others are concerned that this will imperil intellectual practices that have been developed for centuries. Before addressing these claims directly, the chapter will develop its theoretical argument primarily through the use of one type of theory of media ecology: the concept of secondary orality expounded by Walter Ong and, to a lesser extent, his PhD supervisor, Marshall McLuhan.

Ong's seminal work, *Orality and Literacy* identified what he saw as the primary characteristics of oral and literate cultures. Oral cultures, which were the dominant formation until at least the fifteenth-century print revolution in Europe, had the following main characteristics. Memory is the most important aspect of oral cultures as virtually nothing is written down. Because of this, mnemonics in the form of formulaic and repetitious language need to be deployed in order to aid memory recall. It is difficult to store knowledge that cannot be written down, so in oral cultures it must be shared rather than remain private. Finally, there are few means by which accurate information can be disseminated widely and therefore knowledge tends to be restricted to particular locales (it is *subjective* rather than *objective*) (Ong 1982). The use of images was an important characteristic of oral cultures too, vividly demonstrated by the example of religious images in medieval churches reminding illiterate parishioners why they were there (Yates 1966: 50–81). Conversely, in literate cultures the very fact that virtually everything is recorded in writing means that the need to recall information from memory is not as important a characteristic. Language therefore does not need to employ mnemonics and is free to develop in a more abstract way. This tendency is strengthened by the capacity to store information in written form, which enables the development of private, analytical and logical thought. This develops forms of knowledge that are not tied to particular locales, in other words knowledge that tends to be *objective* rather than *subjective*. The main carrier of information in literate cultures, the book, both exemplifies and reinforces these characteristics in its self-containment, capacity to be read privately and silently, and the ease with which it can be disgorged in great quantities far and wide. Until recently, most books over the last few centuries have contained few illustrations; an indication of the marginal role of imagery in literate cultures (Ong 1982).

Though they were writing long before the World Wide Web came into being, both Ong (1982) and Marshall McLuhan (1964) thought that the growth of what they termed electric or electronic media would reach a point where it would begin to significantly reconfigure the existing media ecology. Their basic argument was that these new forms of media contained elements of both oral and literate cultures, and this very fact would need to be incorporated into any future understanding of our so-called literate culture. Ong (1982: 3) coined the phrase 'secondary orality' as a means of conceptualizing these changes. While these ideas did not gain much traction in academia at the time, subsequent developments in digital media have led to a renaissance in the study of the work of both theorists, McLuhan in particular (Lister *et al.* 2003; Van Loon 2008). And this is understandable when one considers how what appear to be residual elements of oral culture are present in latter-day digital communication. While everything is written down and recorded in a digital media environment, unlike literate cultures our capacity for memorization is important as a means of dealing with information overload and as a way of trying to remember where we are and where we would like to go next as we navigate online. The capacity to record all this information in large databases should, like in literate cultures, enable us to develop private opinions, and analytical and logical thought. However, our attempt to make sense of all this information is hindered by the incoherent way in which it is stored, the problems engendered by information overload and – as is particularly the case with social media – much information is stored communally, similar to the way in which it would be in oral cultures. Given that information is not tied to one's immediate locale, then it encourages the construction of *objective* knowledge. However, digital media platforms – like blogs, micro-blogs, message-boards, texting and social networking generally – are designed in such a way as to elicit opinions that are more personal or *subjective*. While the book still retains its self-containment, online texts have much looser boundaries and our attention is constantly drawn away from them by emails, text messages and the general information bazaar that is the World Wide Web. Images are an integral part of the online reading experience, a characteristic of oral societies, rather than the literate societies in which we are supposed to live.

Whether or not we ascribe secondary orality to the societies in which we live, as argued throughout, it does appear that the proliferation of digital media is, however subtly, changing the way in which we view the world. For many, the effect that these changes are having on our capacity to read is overwhelmingly negative. In particular, it is argued that online reading encourages acontextual fact-finding rather than the placing of information within the type of conceptual framework from which knowledge can be acquired. Online readers are said to skim texts rather than parse them, and are constantly distracted by the spectacle that is multimedia (Birkerts 1994; Carr 2010; Greenfield 2006). Some of these writers even go so far as to suggest that these new forms of digital literacy are altering the plasticity of our brains, a literal extension of older arguments about the media 'brainwashing' us (Carr 2010; Greenfield 2006). At this point I do not want to offer an assessment of the validity of these views; instead my answer to these claims will unfold over the latter part of this chapter as I explore online reading practices more comprehensively. In order to do that, I must also engage with the ideas of those who think that digital media's undermining of traditional literary practices such as those described above is to be welcomed, an argument to which I will now turn.

Reading hypertext

Voices from within the academic fields of literary theory and philosophy were critiquing traditional conceptions of literacy long before digital media became as widespread as it is today. In the 1960s and 1970s, the work of French post-structuralists Roland Barthes and Jacques Derrida set about undermining the notion of authorship as a platform from which writers could 'inform' a 'passive' readership (to paraphrase a famous Barthes essay in the late 1960s, they wanted to will 'the death of the author'). Alongside this undermining of the author was a championing of the autonomy of the reader (and of a 'writerly' text whose meaning can be altered in the very act of reading) and arguments for the fluidity of texts (the notion that texts are open not closed, and linked to all other texts rather than being self-contained entities), encapsulated in Derrida's term 'inter-textuality'. All these concepts – death of the author, writerly text and inter-textuality – seem to have more in common with digital texts than the kind of texts that Derrida and Barthes were writing about and it was therefore not surprising that many of the theories that were developed in the early 1990s to conceptualize the rise in hypertext reading contained traces of these writers' ideas (White 2007):

> when designers of computer software examine the pages of *Glas* or *On Grammatology*, they encounter a digitalized, hypertextual Derrida; and when literary theorists examine *Literary Machines*, they encounter a deconstructionist or poststructuralist Nelson. These shocks of recognition can occur because over the past decade critical theory and computer hypertext, apparently unconnected areas of inquiry, have increasingly converged [...] working often, but not always, in ignorance of each other; writers in these areas offer

> evidence that provide us a way into the contemporary *episteme* in the midst of major changes.
>
> (Landow 1992: 2; cited in Pasquali 2014: 42–43)

Landow was joined in making these arguments by Michael Joyce, Jay David Bolter, Peter Lunenfeld and others, some of whom designed as well as wrote about elaborate hypertext stories and academic websites (White 2007).

The proselytizing of new (hypertext) forms of story-telling was carried out with great gusto by a small number of these writers in the early 1990s. While it is not easy to find free versions of these hypertext novels still in existence, a sample of one of the canonical works in this genre, Stuart Moulthrop's *Victory Garden*, can be viewed online (Moulthrop 1995).[10] This is useful in that it enabled the author to compare his reading experience with that of some of the more prominent critics of this genre. One of those critics mentioned above pointedly emphasized how unimpressed he was with Moulthrop's hypertext (Birkerts 1994: 151). While I was only able to access a sampler of the story, I too found the interactive element of the storyline largely irritating and pointless, after a while I was aimlessly clicking on links and reading no more than cursorily. Regardless of whether or not I am convinced that literary conventions in online reading in their totality are being undermined, I did find myself lamenting the absence of narrative coherence in *Victory Garden*. My experience bears out Charney's (1994; cited in Miall and Dobson 2001) contention that hypertext reading places an intolerable burden on readers' short-term memories (as proponents of Ong's concept of secondary orality would expect). Miall and Dobson's (2001) survey of hypertext readers, which showed that 75 per cent of readers felt that they could not adequately follow the narrative in contradistinction to the 10 per cent of readers of the same story in a linear format who felt the same way, seems to validate my subjective impression. This is consistent with two more recent studies which show that the more interactive and less linear an e-book is, the less easy it is for children to follow a story's narrative and read without distraction (Chiong *et al.* 2012), and that readers' desire for interactivity in hyper-fiction is tempered by their requirement that this should not be at the expense of a clear, coherent narrative (Pope 2013).

In order to better understand how disorientating readers find online platforms, it is worth thinking about what it is that they actually value about reading. While many of the opportunities for interactivity between text and reader that digital media has enabled has undoubted benefits, Umberto Eco's (2006: 14–15) argument that readers primarily want to be enlightened by the author, have some sort of guidance as they navigate their way through text(s) and a coherent narrative seems to be validated by studies of hypertext users (Pope 2013). It can also be added that, alongside the continuing popularity of books in their codex form – think of the number of books that an online retailer like *Amazon* sells – is the increasing tendency of digital texts to resemble traditional formats. Indeed, think of the way in which the most successful e-readers very closely replicate original versions of books, even in many cases down to using the original page numbers. The most successful academic publishers use a PDF format for online articles which is virtually identical to the format in the hard copy version of their journals. Aarseth (1997: 74–75) comments that:

> It is revealing and refreshing to observe how flexible and dynamic a book printed on paper can be, and this gives us an important clue to the emergence of digital text forms: new media do not appear in opposition to the old but as emulators of features and functions that are already invented. It is the development and evolution of codex and print forms, not their lack of flexibility, that make digital texts possible.

The difficulty of theorizing digital media use!

The views of 1990s hypertext theorists who believed that online texts would revolutionize reading for the better and of those who are concerned that reading such texts is undermining traditional literacy would seem to have been confounded by studies of contemporary modes of reading. While Chapter 1 highlighted the way in which online reading practices in universities are causing concern among some academics, there is just as much evidence to suggest that students are still able to differentiate between academic and non-academic conventions. What this chapter has illustrated is the complexity of attempting to construct a global theory to explain such a varied range of practices. The three-step model that was outlined earlier is the nearest this author can get in trying to incorporate all the many different elements and platforms of digital media into one over-arching framework.

- STEP 1
- ENFRAMING – orders (structures & commands) a sphere of action (*Technology*) +

- **STEP 2**
- BINDING-USE – technology is only accepted once it becomes bound to a particular use (*User*)
- = SUCCESSFUL TECHNOLOGY whose use is restricted by its enframing (**STEP 3**)

This model is flexible enough to incorporate the media ecology theories of Neil Postman and Walter Ong; indeed, the three-step model itself should be viewed as an ecological model. In relation to online reading in particular, Walter Ong's concept of secondary orality is a compelling explanation for contemporary practices. Ong's (1982: 136–137) argument is that secondary orality differs from its primary antecedent in that it is to an extent freely chosen – pre-literate cultures could not *choose* between orality and literacy. Our apparently seamless movement across myriad reading platforms, digital and otherwise, suggests that Ong's work is correct in its emphasis on user agency. But, as the three-step model proposes, enframing does limit our autonomy in relation to use and, as standardization of many platforms continues apace, we should not over-emphasize the power of individual agency.

A lot of this theoretical discussion has left more questions than answers. This author's refusal to indulge in futurology means that I will not follow much bolder writers who try to find those answers in contemplation of 'future trends'. I believe that a more fruitful avenue is to discuss these conceptual issues within the context of a wide range of present practices. To that end, the next three chapters will analyse the use of digital media in new forms of social and political activism, surveillance and in the developing world. While these analyses will reinforce the sense of the complexity of theorizing these interactions, in its highlighting of a varied range of practices it is to be hoped that it will provide some useful insights into the relationship between digital media and social practices.

7
The New Social Movements

Introduction

Chapter 3 made an attempt to establish a normative model for a digital-media-infused politics in liberal democratic societies. While this theoretical framework can be used for most types of political activity – in this sense it can be used for the types of activity that will be discussed in this chapter – it is most credibly applied to electoral politics in liberal democratic societies. In order to extend some of the ideas in Chapter 3 to a study of extra-parliamentary political activities, the work of theorists like Michael Hardt and Antonio Negri will be used to augment my normative model. In doing so, the chapter focuses mainly on political activities and social campaigns that take place largely outside the mainstream political arena. These will be mainly associated with the new social movements, but will also include Barack Obama's 2008 presidential campaign; this was chosen as a case study here precisely because of its difference from most campaigns in formal elections.

Digital media has always been associated with social change. While many of the early pioneers of the Internet were motivated by an antipathy towards conventional politics, they were not apolitical. Indeed, it could be argued that the vision of society being transformed by computerization that emerged from the late 1960s Californian counterculture movement constituted the first social movement of the burgeoning digital media age. It was largely successful too, as the so-called 'Californian ideology'[11] that they espoused was a type of libertarian politics that infuses debates about the mission of the World Wide Web even today. In addition to the value of discussing the Californian ideology as a successful social movement, its libertarian politics makes it a useful counterpoint to the mainly left-leaning new social movements on which this chapter will mainly focus, and is therefore an appropriate place to start.

The Californian ideology and the growth of the Internet

Before discussing the role of the Californian counterculture in developing the Internet, it should be stressed that the impulse of the original designers was a military one. The launch by the Soviet Union of the *Sputnik* satellite in 1957 signalled to the USA that it needed to take a drastic action to avoid falling significantly behind in the race to develop the newest military technologies. Its initial response was to establish ARPA (Advanced Research Projects Agency) as a means of giving the nation's leading scientific researchers the freedom to develop innovative ideas to counter this threat. Under the influence of John F. Ford, the CIA's main expert in information technologies, ARPA turned its attention in the early 1960s to the development of communication networks; as early as 1956 the Soviet Union had built a computer network to coordinate Moscow's air defences (Barbrook 2007: 150–152). While scientists in a small number of elite institutions were required to carry out military research in an era of US history characterized by anxiety, paranoia and a culture of surveillance, this kind of atmosphere tends not to encourage innovative research. Therefore, in contrast to working practices in much of American society at that time, scientists were given almost *carte blanche* to develop their own ideas, the result of which was that some could also experiment with projects outside the parameters of ARPA's prescribed agenda (Chadwick 2006: 39–40; Rheingold 2002: 210). Its first director (who served from 1962 to 1964), J.C.R. Licklider, certainly envisioned a future of networked computerized communication systems that exceeded the limited, military-inflected scope of ARPA's mission. In 1960 Licklider had written a paper entitled 'Man-Machine Symbiosis' in which he speculated on how computers could be used to enhance human knowledge through the use of networks of information. His position, though short-lived, as director of ARPA, enabled him to create a free-thinking culture in which ideas about information networks could flourish (Chadwick 2006: 39–40; Rheingold 2002: 211). This resulted in ARPA's funding of the Augmentation Research Group (ARC) at the Stanford Research Institute, which was headed by another prominent visionary, Douglas Englebart (Turner 2006: 108). Both Licklider and Englebart were heavily influenced by Vannevar Bush's famous 1945 article, 'As we may think'. In it, Bush speculated about the possibility of constructing a 'memex', a machine in which people could search for micro-film records that had been indexed in the associative way that he believed the human mind worked (Rheingold 2002: 211; Turner 2006: 106–107). When Engelbart read Bush's work just after the Second World War had ended, he found himself in sympathy with the author's view that technologies, like nuclear weapons, which could destroy humankind, needed to be augmented by those which could raise human knowledge and consciousness to a new level (Turner 2006: 107). These ideas eventually led to ARC's development of a prototype Internet in the late 1960s, called the OLS (Online System) or NLS (Turner 2006: 107–108; Wessels 2010: 20). The race to build what we now call the Internet had become official and was actualized in the creation in 1969 of ARPANET, the computer network commonly considered to be its first proper iteration.

The 1970s was characterized by a raft of innovations that built on this backbone, most notably Vint Cerf and Robert Kahn's TCP/IP protocols early in that decade. However, despite the free-wheeling nature of work in this field, these innovations were being developed largely by elite scientists and computer programmers within a relatively closed system of elite universities, scientific institutions and military establishments. However, as the decade unfolded, a confluence of a growing breed of computer 'hackers' – technicians devoted to free experimentation outside institutional parameters – within these organizations, the development of personal computers (especially at Apple) and the invention of the modem, a technical device that linked individual computers to the network, expanded the sites of innovation. Various unofficial networks flourished and the first online bulletin boards appeared in this decade (Chadwick 2006: 43). One of the earliest and most successful online communities in the next decade, 1980s, was the WELL (Whole Earth 'Lectronic Link). Established in 1985, the WELL was not only a technical success and of great benefit to its few thousand users, but also provided a platform for its founders to spread their gospel at a critical stage in the history of the Internet, particularly as the World Wide Web hoved into view:

In 1985 he [Stewart Brand] gathered them again on what would become perhaps the most influential computer conferencing system of the decade, the Whole Earth 'Lectronic Link, or the WELL. Throughout the late 1980s and early 1990s, Brand and other members of the network, including Kevin Kelly, Howard Rheingold, Esther Dyson, and John Perry Barlow, became some of the most-quoted spokespeople for a countercultural vision of the Internet. In 1993 all would help to create the magazine that, more than any other, depicted the emerging digital world in revolutionary terms: *Wired.*

(Turner 2006: 3)

Wired's politics, especially in relation to the digital economy, have been discussed elsewhere in this book and do not require much elaboration here. Some accounts of the development of the Internet argue that the magazine 'hijacked' what was essentially a 'New Left' countercultural agenda (Barbrook 2007: 264). However, much in the same way as it is not easy to disentangle the two supposedly different strands of the Internet's development – the defence establishment's research into communication networks for military purposes and the more egalitarian vision of people like Engelbart and Licklider – the motivations of the counterculture were never that far removed from the mainstream society that they had turned against. Whether or not it is true that the primary reason for developing the Internet was to create a communication network that could withstand a nuclear attack, the military and the counterculture both wanted to build the same kind of global communication system. As Fred Turner (2006: 76–78) has expertly documented, in many of the countercultural communes that mushroomed in the late 1960s and 1970s in various parts of America, gender, class and race divisions were just as pronounced as they were in mainstream society. Moreover, the hierarchy that many early Internet pioneers were supposedly breaking down could not be entirely suppressed even within their own communities, even if this was not acknowledged as such (Turner 2006: 64–65). Internet pioneer Stewart Brand's concept of *collective* consciousness was based paradoxically on *individuals* using small technologies, including computers, to defend their personal freedoms (Turner 2006). In this, their vision of the Internet has been compared with the development of citizen band radio in the 1970s which, while it could be seen as having the potential to link local communities, was more usually associated with libertarian politics and consumerism:

> CB radio was rarely used for anything more politically significant than giving speeding truckers the chance to elude the highway patrol, a sort of minimally anarchist gesture of rebellion. For the most part, it became another consumer novelty.

(Roszak 1988: 195–196)

As the anecdote about novelist Ken Kesey's bizarre demand to anti-war protestors at Berkeley in 1965 that they should they should end their campaign because they were too confrontational, even their opposition to militarism could be tempered by their visceral disdain for anything that smacked of conventional politicking (Turner 2006: 64). We have already seen in Chapter 5 that this libertarianism manifested itself mainly in the championing of the deregulation of the digital economy, but the chafing against any restrictions on free expression, best exemplified by John Perry Barlow's *Electronic Frontier Foundation* (EFF), was equally important to them. All the above reinforces Turner's (2006) argument that the counterculture was not so much interested in conventional left-wing politics as it was in libertarianism.

An acknowledgement of the primacy of the libertarianism in the politics of the Californian ideology paradoxically highlights how successful the movement has been in spreading its ideas into mainstream society, for it is less the case that its aims were diluted when they seeped into mainstream society and more that those aims were not particularly left-wing in the first place. As the discussions in the first two chapters about the freedom to build knowledge without the mediation of educational institutions and even to create our own identities, and in Chapter 5 about the so-called new economy all demonstrate, this strain of libertarianism is the semi-official ideology of digital media. As this chapter and the next will go on to show, particularly in relation to some forms of hacktivism and sensitivity to government surveillance, this even extends to activism of a more left-wing bent. If we define a successful ideology as one which forces even its political opponents to argue on its terrain, then the Californian ideology can be thus described. Its capacity to effect social change without diluting its aims thus gives some hope to latter-day social movements, notwithstanding the fact that many of them are not of the same political ilk as the former. How they might go about doing this is the subject of the next section.

What are the new social movements?

Earlier in this chapter, we saw how disillusionment with conventional political ideas and the formal political process led a small group of counterculture computer hackers to create early computer networks in their own image, a project so successful that their ideas continue to have disproportionate influence to this day. Similar thinking was taking place on the Left too, where the disillusionment with totalizing ideologies like Marxism – which subordinated all political issues to the class struggle and revolution that never looked like it was coming – became pronounced during the 1960s and 1970s as more people began to grapple with philosophies that would explain the emerging struggles centred on race, gender and sexuality in western society. It was in this context that the concept of 'social movements' emerged in the 1970s.

> The idea of social movement was conceived, at least in my mind, in opposition to the traditional concept of class conflict. Not opposition in the sense of being reformist. Instead, when we speak about class conflict we refer, basically, to a process of capitalist development or a process of social and economic crisis in objective terms. When we began speaking, a long time ago now, about social movements, we tried to elaborate a new approach and to pass on the actors' side.
>
> (Touraine 2002: 89; cited in Jordan and Taylor 2004: 46–47)

The growth of identity politics in the 1970s has been oft-told and does not need recapitulation here. But it is worth noting that the rise of single-issue politics was arguably fuelled by the disillusionment with global theories like Marxism brought about by seminal events like the Soviet Union's invasion of Czechoslovakia and the failure of protesting Marxist students in France in that same year, 1968, to effect radical change. The critique of Marxism was extended to all theoretical and political meta-narratives by, among others, a group of highly influential, mainly French, post-structuralists. In this worldview, the most effective political strategies were those that focused on the 'local' or on particularistic identities. (For this reason, these social movements differed from the Californian ideology in the latter's optimistic belief that it was still possible to change society in its totality.)

Like the Californian counterculture, social movements used nascent computer networks to put their politics in practice and to disseminate their ideas to a wider, if limited, audience. One of the most well-known and successful examples of this type of activism was the use of the Amsterdam free radio's adoption of digital technologies in the late 1980s to advance its social and political goals (Lovink 2011: 126). This eventually led to the creation of the Amsterdam Digital City in 1994, a project which facilitated the discussion of a whole range of social and political issues within designated virtual zones (Curran 2012: 39). Also, like the Californian counterculture, the development of the World Wide Web enabled initiatives like these, which were essentially local, to 'scale-up' and have influence at a global level.

A famous example of this is what many regard as the first significant campaign by new social movements, the popularizing of the ideas of the Zapatistas.[12] On New Year's Day in 1994, an indigenous Indian army called the EZLN (more commonly known as the Zapatistas) captured a number of towns in the southern-most state of Mexico, Chiapas. The same day marked the coming into effect of NAFTA (North American Free Trade Agreement), a free trade treaty which, they argued, would imperil the land rights of indigenous people in Mexico and, in and of itself would be a reason for activists in the region to oppose it. But the Zapatistas were also opposed to the growth of neo-liberalism globally and this, along with a mysterious and charismatic leader, Subcommander Marcos, enabled them to gain significant worldwide support. But the initial impact of their campaign arguably would not have been as successful without the use of La Neta, a computer network developed over the previous four years with the help of the Institute for Global Communications in San Francisco, a wide range of women's non-governmental organizations (NGOs) and the Catholic Church, which enabled local groups in the Chiapas region to get online and spread their message before the insurrection took place (Chadwick 2006: 126). Thus, while the insurrection itself was quickly and brutally crushed by the Mexican army, global support for the Zapatistas' aims grew exponentially as their message spread rapidly through transnational computer networks. The Zapatistas had a wider and longer-lasting effect, in that they inspired activists from around the world to set up networks to challenge neo-liberalism, creating an infrastructure which eventually led to the anti-World Trade Organization (WTO) protests at Seattle in 1999 (Juris 2008: 45).

Towards the end of the 1990s, these Zapatista-inspired new social movements thus became even more active, chalking up a major success when the proposed multi-lateral agreement on investment (MAI), drawn up by some of the leading developed nations to give multinational corporations a whole range of exemptions from national laws and regulations, was struck down in 1998 after a major email campaign (Sarikakis and Thussu 2006: 7). The WTO summit at Seattle in 1999 was met with a massive protest, with campaigners' use of digital media both as a means of coordinating the movements of the protestors and to disseminate their message confounding the American authorities. The most significant characteristic of all these movements was that each featured a myriad of social organizations with differential goals, coalescing around one specific campaign. In this sense, they appeared to shy away from the meta-narratives of traditional forms of politics while also, in their participation in the formation of broader alliances, differing from the traditional social movements in their transcending of single-issue politics. But how might a progressive political movement be fashioned out of these myriad campaigns without employing a meta-narrative? The most credible attempt at answering this question has come from Michael Hardt and Antonio Negri's formulation of the *Multitude*, to which I will now turn.

Hardt and Negri's multitude

Castells's earlier cited argument that the global network forces politics and social activism to operate under its own logic has led some to argue that progressive forms of social change cannot be fashioned through heavily commercial and deregulated digital media platforms (Mejias 2012). But, as Hardt and Negri (2006) – two theorists with hardly a less critical take on the role of capitalism in the development of global computer networks than those who are sceptical of their potential for promoting progressive social change – argue, it is simply not possible to turn the clock back to a time when our communication technologies were less advanced. Indeed, the immaterial labour that global elites – or what Hardt and Negri call the 'Empire' – utilize has simultaneously an immanent potential for liberation (Dyer-Witheford 2007: 200). Given that any contemporary progressive political project cannot avoid using the media if it wants to promote its cause effectively, then the emphasis must be on promoting types of digital media activities and platforms which have some sort of popular, democratic control. The fact that, as Castells and Harvey argued in Chapter 4, political and financial elites – Hardt and Negri's Empire – have managed to conquer space at a global level means that this new political project cannot be confined to individual national territories. Hardt and Negri (2006) have argued that this type of activism is already happening in what they term the 'multitude', a body (albeit fluid) of progressive people organized in a non-hierarchical way on the basis of what they have in common. It might seem to be a truism that people should organize on the basis of what they have in common, but Hardt and Negri use this formulation to stress the importance of disparate groups putting their differences to one side in order to pursue a specific political project that is similar to, but subtly distinct from, the pre-capitalist conceptual space termed 'the commons' (Hardt and Negri 2006: xv). They conceptualize these people as 'singularities' who, though globally dispersed, form the multitude, but, importantly, without ridding themselves of the opinions they hold which differ from those of the other members. Examples of multitudes include the alter-globalization[13] movement and the opposition to the invasion of Iraq in 2003, both of which include disparate groups like anarchists and Christian organizations who profoundly disagree on many other political issues. And while this form of network is not synonymous with Internet networks, they argue that

> Once again, a distributed network such as the Internet is a good initial image or model for the multitude because, first, the various nodes remain different but all are connected in the Web, and, second, the external boundaries of the network are open such that new nodes and new relationships can always be added.
>
> (Hardt and Negri 2006: xv)

In this vein, they praise the role of global activist website Indymedia,[14] which was first created for the 1999 anti-WTO Seattle protests, but today operates across a much wider political terrain (Hardt and Negri 2006: 305). The role of the type of alternative media platform that Indymedia represents in promoting the ideas and campaigns of new social movements will be discussed in more detail in the account of the anti-G20 protest in London in 2009 below.

Other activist writers too have invested faith in new communication technologies as a means of promoting an alternative to global capitalism as it is presently constituted; Castells also writes about the potential of the network to do this in relation to the global environmentalist movement (Castells 2010b: 168–191). Naomi Klein (2000: 396) argues that:

> the Net is more than an organizing tool – it has become an organizing model, a blueprint for decentralized but cooperative decision making. It facilitates the process of information sharing to such a degree that many groups can work in concert with one another without the need to achieve monolithic consensus … [Hardt and Negri's idea of the singularities making up the multitude].

Hardt and Negri (2006: 185–186) themselves are a little more sceptical, lamenting the publicly-oriented Internet's metamorphosis into the more commercially-oriented and privatized World Wide Web. Nonetheless, it is clear that digital media enables social movements to both scale-up local or particular projects to a global level and to more quickly link with others on single-issue campaigns. In their own words: 'It takes a network to fight a network' (Hardt and Negri 2006: 58).

Hardt and Negri's concept of the multitude can be considered in relation to a number of contemporary social movements, two of which will be analysed below. On 1 April 2009, protestors in London demonstrated against a summit of G20 leaders held in the same city. As one of the first opportunities for international social movements to gather in one of the world's leading financial centres since the previous year's bailout of the global banking sector, this march was much anticipated by activists. As well as providing a useful case study of the multitude in action, the march became notorious as a result of the death of a bystander who had been violently pushed to the ground by a police officer. The subsequent coverage of this tragic event illustrated the growing importance of alternative forms of media in countering mainstream narratives and, as far as this chapter is concerned, will trigger a discussion on the use of tactical media by the multitude. Earlier in the same year, a relatively novice politician, Barack Obama, became the first African-American president of the USA. His defeat of a seasoned campaigner like Hillary Clinton for the Democratic party nomination and subsequent victory in the presidential race itself, despite being a politician who even five years previously was unknown to the vast majority of the electorate, invited intense scrutiny of his campaigning methods. Obama's campaign was characterized by the huge number of volunteers, many of whom came from demographics considered hitherto to be not that interested in electoral politics. Coordinating all their activities and disseminating Obama's message far and wide was carried out with the skilful use of various forms of digital media. While it would be reductive to claim, as many have, that it was the Internet that was primarily responsible for Obama's victory (Stirland 2008), his campaign can be considered an example of a successful social movement. Given that the Obama campaign represents an uncommon case, in that it was a social movement wedded to electoral politics, the chapter will address these two cases in reverse chronology and thus will begin with the 2009 G20 demonstration in London.

2009 G20 protest in London

The 2009 London meeting of G20 leaders to coordinate an international response to the continuing global financial crisis provided social movements critical of the role of capitalism in triggering and sustaining the downturn an opportunity to publicize their cause. Unlike previous summits, where the organizers were surprised by the scale of the protests, the mainstream media and UK authorities were fully prepared for the G20 demonstration in 2009. One of the tactics that world leaders adopted after the 1999 Seattle protests was to hold meetings in remote places which were not easily accessible to activists. While this tactic was not employed in this particular case, others were, including the spreading of disinformation through digital media and the largely compliant mainstream press. Thus, despite UK newspapers' critical stance towards the irresponsibility of the banking sector and the subsequent bailout, most reports in the run up to the demonstration highlighted the supposed threat of violence from protestors (Hands 2011: 153). Even left-leaning newspapers tended to adopt this line, as demonstrated by these lurid speculations in *The Observer*:

> Details of direct action, gleaned from chatter on anarchist websites and meetings attended by the Observer, include a rumoured plan to block the Blackwall Tunnel and cause a security scare on the London Underground by leaving bags unattended on trains. There is also speculation that protesters will drive a tank to the ExCeL conference centre in London's Docklands, where the G20 are meeting, and attempt to harass politicians with wake-up calls to their hotels in the middle of the night. None of the organisers of the peaceful demonstrations say they are aware of any such tactics.
>
> There are growing fears for the safety of people making their way to work on 1 and 2 April. A spokesman for the London Chamber of Commerce said: 'Businesses might want to consider asking their staff not to dress in a suit and tie as a lot of the protesters say they're going to target bankers…'.
>
> (Smith and Rogers 2009)

As well as accepting uncritically the views of the police, newspapers drew on two other main sources: the words of Chris Knight, a professor of anthropology at the University of East London, and pronouncements on a website entitled *G20 Meltdown* (Hands 2011: 152–154).[15] The *G20 Meltdown* website featured images of the four horsemen of the apocalypse, each of which would lead a march on the day of protest. The website also distributed flyers, including one which promised to 'storm the banks'. Knight was identified by the media as a leading figure in the *G20 Meltdown* organization and both he and it were tainted by his comments in an interview on BBC Radio 4 a week before the march, in which he speculated that there might be 'real bankers hanging from lampposts' (quoted in Hands 2011: 152). What actually transpired on the day were some forceful policing manoeuvres, including the 'kettling', or penning in, of large groups of protestors for hours at a time, culminating in an assault on a bystander, Ian Tomlinson, who subsequently died. While the UK press might have prepared the ground for this robust policing in its over-zealous reporting of protestors' alleged intentions, any attempt to exonerate the police crumbled in the days that followed, as footage from various digital devices showed that the protestors were responsible for little of the violence that took place on the day (Hands 2011: 154; Hill 2013: 32).

In the immediate aftermath of the march, Indymedia in particular became a hub for eye-witness statements from demonstrators, and photographs and footage of police violence. Indeed, on the very next day after Tomlinson's death, the mainstream media narrative of a mob of people preventing the emergency services from dealing with the stricken man was being challenged on Indymedia London's website in a video testimony of two people who witnessed the aftermath of the assault (Indymedia London 2009).[16] On 7 April, less than one week after his death, the *Guardian* uploaded onto its website a short video taken by, like the victim, someone who was not involved in the protest, which showed Tomlinson being shoved to the ground by a police officer shortly before his collapse and subsequent death (Lewis 2009). In one respect, this can be read as consistent with the stance that the author of the article had already adopted as one of the few journalists to question the attitude of the police in the lead-up to the demonstration (Lewis, Laville and Vidal 2009). But it could also be viewed as an indication of the increasing realization by the mainstream media that merely parroting official statements as a means of establishing the narrative of incidents like these exposes it to potential ridicule. The pressure to trust the credibility of these official statements emanates not only from the worldview of newspaper proprietors, but also in the UK at least from the 1952 Defamation Act which protects journalists from being prosecuted for libel if their information is taken from official sources: 'In short, if an official statement accuses a man of being a criminal, the [UK] press can't be sued for using it; but if a man accuses a government official of being a criminal, the press can be sued for repeating it' (Davies 2008: 122). For this reason, even the most left-leaning newspapers have tended to proceed with caution in challenging the police's version of controversial incidents, but now the easy availability of footage refuting the official line means that the press, if it so wishes, can be much more critical in its reporting. *The Guardian* and the BBC, both of whose websites willingly hosted footage that challenged the police's view of events during the G20 march, have been at the forefront at this new form of reporting (Couldry 2012: 24). So, while as figures from 2008 show, Indymedia was only the 36,694th world's most visited website (the BBC, CNN and

New York Times were 44th, 52nd and 115th, respectively (alexa.com figures cited in Fenton 2012: 135)), the willingness of some mainstream news websites to use its resources gives it an influence way beyond its modest exposure. This trend has continued since 2009, especially with *The Guardian* newspaper, whose work with Wikileaks and NSA (National Security Agency) whistleblower Edward Snowden has given credibility and wider coverage to revelations that might otherwise have been relegated to the margins of public discourse. In the case of Ian Tomlinson's death, the police officer, PC Simon Harwood, who shoved him to the ground was quickly identified and brought to trial in 2012, where he was acquitted of manslaughter; Harwood was, though, dismissed by the Metropolitan police for gross misconduct in September of the same year (*BBC News* 2012). So, while this case thus demonstrated the effectiveness of digital media in identifying that Tomlinson had been struck by a police officer, the absence of a conviction for the police officer involved suggests that the treatment of the police by the law courts has not changed much since the days when these incidents were routinely unrecorded.

Despite attempts by the mainstream media to suggest otherwise in their focus on Chris Knight and the G20 website, in the wide variety of organizations involved on the day this was an example of the multitude in practice (Hands 2011: 154). Their use of a wide range of digital media, both to coordinate their activities and disseminate their message, was partially effective, even in the teeth of a largely hostile mainstream media. Paradoxically, though critical of the actions of the Metropolitan Police on the day, the mainstream media's focus in the aftermath of the demonstration on the tragic death of Ian Tomlinson marginalized discussion of the wider aims of the protestors in the days and weeks that followed the march. This problem was identified by Jordan and Taylor (2004: 152) long before this particular march took place, in their observation that the global media is interested in 'vertical' events or spectacles, the coverage of which tends to eclipse analysis of the 'horizontal' form of politics practised by the multitude.

The 2008 USA presidential election

Another example of a social movement which attracted the global media's attention was Barack Obama's successful US presidential campaign in 2008. The appropriateness of calling Obama's campaign a social movement can be defended on the grounds that it drew together a huge number of people from disparate backgrounds, many of whom had shown little interest in electoral politics before that. The bald figures are staggering. His campaign manager, David Plouffe, claims that Obama had 13 million people on his email list, four million contributors (Stirland (2008) puts this figure as over three million raising a staggering $600 million) and six million volunteers (Hands 2011: 115). The central node of his campaign was the MyBo website, which had been set up by Facebook co-founder, Chris Hughes. On this website alone, there were more than 1.5–2 million profiles, the coordination of 150–200 thousand events, formation of 35,000 groups, and the creation of 400,000 blogs and 70,000 personal fund raising pages from which $30 million was raised. The campaign even had its own dedicated YouTube channel, in which 1,800 videos were created with 110 million views up to election day (Hands 2011: 115–116; Hill, S. 2009: 11; McGirt 2009; Stirland 2008).

In attracting followers who had an activist bent, Obama was treading a path travelled by Howard Dean, a Democratic candidate who did well in the primaries in 2004 without winning the party's nomination. While Dean exploited the online payment systems that people had become so familiar with as an efficient means of receiving numerous small donations, the candidate's personality and, by extension, the type of voter he was trying to attract, helped him too:

> Not everyone can do what Dean has done. The Internet plays to his strengths – just as TV played to the strengths of John F. Kennedy. Carol Darr, director of the Institute for Politics, Democracy & the Internet in Washington, theorizes that 'charismatic, outspoken mavericks' are the ones who attract Internet followings. 'Since the Internet is interactive and it requires the user to take an affirmative action, to go to a Web site, to log on to a chat room, you have to have candidates that motivate people,' she says. 'It's not like couch potatoes, where you just sit there and are the passive recipient of whatever commercial comes along.'
>
> (Mack 2004; cited in Barron 2008: 8)

Earlier so-called maverick politicians had exploited the new technologies, if not to the same extent as Dean: in 1998 former professional wrestler Jesse Ventura surprisingly won the race to become Minnesota's governor, a success partly attributed to his emailing 4,000 potential supporters (Barron 2008: 4). Another politician considered to be outside his party's mainstream, John McCain, was also an early adopter of digital media, as this observation from his campaign for the Republican presidential nomination in 2000 testifies:

Here's how politics has invaded cyberspace: In the hours of darkness between John McCain's victory in New Hampshire and the start of the next business day, he raised $300,000 and found 4,000 new volunteers through his Website. By week's end the total was $2 million and 22,000 volunteers – numbers that would have been even higher had the site not become temporarily dysfunctional because of heavy traffic.

(Birnbaum 2000: 84; cited in Barron 2008: 4)

Why, then, was McCain not able to more fully exploit his expertise in Internet campaigning against Obama during the 2008 presidential race? A superficial answer to the question would be that, as McCain and Dean's unsuccessful nomination campaigns in 2000 and 2004 showed, mastery of digital media technologies does not in and of itself guarantee success. Against that, it could be argued that both candidates would not have gone so far without these techniques and that McCain actually did succeed in 2008 to the extent that he won the Republican nomination. Also, their respective campaigns were nowhere near as sophisticated as Obama's operation, particularly in its capacity to get supporters involved in activities beyond donating money and lending their votes. Obama's campaign was especially successful at garnering the enthusiasm of young people, engaging them on their social networking profiles rather than simply directing all communication to MyBo. While McCain made a valiant attempt late in the day to connect with voters through these social networking websites, he was still outgunned by Obama by around five to one in terms of Facebook supporters, six to one on MySpace and he lagged behind in YouTube channel subscribers by a ratio of 11 to one (Pew Research Center's Project for Excellence in Journalism 2008: 1–2). In what many people regard as the 'real' presidential race, the battle for the Democratic nomination between Obama and Hillary Clinton, the former Senator for New York's style of communication on the Internet was characterized by a clumsiness that made her look out of touch with voters, especially younger ones (Hirschorn 2008; cited in Barron 2008: 12–13). Barron (2008: 20) cites a *New York Times* post-election survey, which shows that 66 per cent of the 18–29 year olds who voted chose Obama, an indication of the success of his campaign strategy in reaching out to this demographic via social networking websites. While US voters are still mainly 'white' and middle-class, Obama's campaign was brilliant in bringing out the vote of the growing African-American and Hispanic electorate, especially those in the 18–24 years age range (File 2013). There was also a better turnout of lower-income voters, though their figures still remained low (Demos 2008). This points to the success of Obama's campaign in enthusing large numbers of voters from a disparate range of backgrounds (remember that he had three to four million contributors and six million volunteers) that hitherto had felt that the formal political process 'was not for them'. His subsequent victory suggests that this is an example of a successful social movement and, in its skilful use of digital media, seemingly a multitude *par excellence*.

Tactical media and hacktivism

Both these examples demonstrate the effectiveness of the 'tactical' use of media. Tactical media was a term which first emerged in the early 1990s to describe the DIY media, some of which was digital, that flourished in the aftermath of the fall of the Berlin Wall (Lovink 2008: 186). This period was characterized by the rapid growth of individual ownership of personal computers and other media, increased access to the Internet and a political climate in which the liberal democratic capitalist state appeared to have no serious challenger. Thus, there was a paradox in that the myriad opportunities for dissent that these new forms of media afforded operated within the parameters of a political system centred on a liberal democratic model which seemed impervious to significant change. This crude empirical fact, as well as the effect of decades of postmodernism in undermining the case for revolutionary change based on grand narratives like Marxism, is acknowledged by theorists of tactical media and shapes practice too (Lovink 2008; Raley 2009). While not explicitly discussing tactical media, Hardt and Negri's (2006) position is similar in their postmodern argument that there is no longer a place outside the system for critique:

> Here is the strong novelty of militancy today: it repeats the virtues of insurrectional action of two hundred years of subversive experience, but at the same time it is linked to a new world, a world that knows no outside. It knows only an inside, a vital and ineluctable participation in the set of social structures, with no possibility of transcending them. This inside is the productive cooperation of mass intellectuality and affective networks, the productivity of postmodern biopolitics.
>
> (Hardt and Negri 2000: 413; cited in Jordan and Taylor 2004: 150)

This does not mean that the system cannot and should not be disrupted and often it is the temporal tactic which can create short-lived freedom from global capitalism's spatial hegemony (Raley 2009: 12; White 2012). The successful campaign against the MAI in 1998 and the Seattle protests a year later are examples of the capacity of digital media to use the speed with which it can coordinate activities and disseminate ideas to wrong-foot slower-moving governments. This tactic was augmented by the sabotage of websites of the organizations which the protestors opposed; examples include the attacks on the WTO's website during the 1999 protests and cyber assaults in support of the Zapatistas (Chadwick 2006: 126–133; Jordan 2007). These disruptions of the digital infrastructure of organizations can be characterized as 'hacktivism', a conflation of the words hacking and activism. While hacking is commonly viewed as a negative, if not downright dangerous activity, it has a long, largely positive, history. Originating in the free-wheeling research environment at MIT in the 1950s, as this chapter has illustrated, most developments in computation since then have been driven by individual experimentation, often from people operating at the margins of business and academia (Lister *et al.* 2003: 265). Hackers were mainly concerned with the technical challenge of building better IT systems, though, as the earlier discussion on the Californian ideology showed, some believed that their work should be in the service of creating an improved society. Despite this, there is a widely held view that hackers' politics do not extend much beyond the desire to make information freely and widely available, that they are most interested in changing online, rather than offline, cultures and that they represent a technical elite rather than a mass movement (Jordan 2007: 74–75; O'Neil 2009: 14–15, 37). For this reason, the contemporary use of tactical media for the pursuance of social and political goals tends to be referred to as hacktivism.

In Jordan's (2007: 79–82) conceptualization of two distinct forms of hacktivism, what he refers to as the 'digitally correct' approach is similar to hacking in that it seeks to 'free' information from its technical and economic restrictions. And, like hacking, it can be charged with advancing the interests of a technical elite, rather than that of mass society. In this sense, Jordan's other iteration, 'mass action hacktivism', is not only more democratic, but also appropriate for the long, hard slog of activism in which new social movements are involved. In the former category are actions like DDoS (Distributed Denial of Service-attacks), which can be carried out by individuals. The activities of the

NSA whistleblower Edward Snowden and, to a certain extent Wikileaks, fall under this rubric. The WTO's website was disrupted around the time of the Seattle protests in 1999 by overloading the server with huge numbers of users. The result of this kind of action is similar to that of DDoS, but, importantly for a mass movement, is more democratic in its use of large numbers of people rather than small or even lone numbers of tech-savvy individuals (Jordan 2007: 75–78). As such, it can be described as mass action hacktivism. While there appears to be a division between those who want to liberate information and those who disable websites, and thus restrict the free expression of the organizations affected, the gap is not always apparent. An example of this was the group *Anonymous's* take-over of PBS's website because the network had aired a critical documentary on Wikileaks's Julian Assange: an attack on free expression in the name of free expression (Halpern 2012)!

The multitude, G20 and Obama

The Obama and the G20 campaign demonstrate the strength of these types of social formations, but also highlight the conceptual flaws in Hardt and Negri's idea. As many writers have pointed out, Hardt and Negri's multitude appears to lack agency, in the sense that it appears to come into being (and then disperse) spontaneously (Cavanagh 2007: 43; Dyer-Witheford 2007: 201–202; Hands 2011: 169, 172). Also, Hardt and Negri are unconcerned about building the type of social institutions which are usually needed to sustain a progressive political project (Couldry 2012: 118). The success of Obama's campaign and the increasing popularity of the ideas of the alter-globalization movement are partly the result of the skilful use of tactical media. Castells argued in Chapter 3 that the logic of the network is to encourage personality politics and, in this sense, Obama's outsider status was of greater benefit to him than it would have been a generation ago. But Castells's point about personality politics can be extended to incorporate all forms of spectacle. This explains the increasingly carnivalesque character of contemporary protests, with demonstrators dressing in elaborate costumes and/or performing political theatre (Juris 2008; Rojecki 2002). And it is not only global social movements that use this tactic, as global communication networks enable national, regional and local movements to scale-up. The Zapatista campaign discussed earlier is one example of this, but others too, like the global attention that Belarusian pillow-fighting activists' highlighting of police repression received in 2010,[17] have used the network to scale-up their protests. But, as the inability to secure a conviction for the police officer who shoved Ian Tomlinson illustrates, it will take more than exposure by digital media to bring about significant social change. (This point is reinforced by subsequent high profile cases involving filmed footage of the Metropolitan Police manhandling student protestors where no disciplinary action has ensued (*BBC News* 2011; Hill 2013: 35; Payne 2013).)

There are factors other than their respective use of tactical media which might explain the effectiveness of the G20 demonstrators and Obama campaigners. The global financial crisis is similar to the fall of the Berlin Wall in that it contains the immanent potential for bringing about a paradigm shift in mainstream thinking about global economics. Our present global financial architecture looks much more precarious than it did a decade ago and, consequently, criticisms of it which back then would have been reduced to the margins of public discourse are now part of mainstream debate. In other words, the increasing popularity of these critiques might have less to do with the effectiveness of the new social movements than the empirical evidence of the faultiness of neo-liberalism through the near collapse of the global banking system. This factor was arguably also one of the most decisive in propelling Obama to the White House. Chapter 5 showed how the bailout of the automobile industry arguably contributed to Obama's re-election in 2012 and the importance of the economy to the voters in 2008 is starkly illustrated by the fact that 63 per cent of voters considered it *the* most important issue (Pew Research Center 2008). In this political climate, it is not surprising that the candidate representing the party of government suffered from the collapse of banking giant Lehman Brothers in mid-September of that year:

According to Pew Research Center public opinion surveys, McCain had trailed Obama for much of 2008, but by the week of Sept. 9–14, in the aftermath of his convention, he had pulled even with Obama among likely voters. In every Pew Research Center survey taken after the financial crisis of mid-September, McCain trailed Obama by no less than six points. And in one particularly telling poll taken within a week of the Lehman Bros. bankruptcy, fully 30% of those surveyed said their opinion of McCain had become less favorable in the past few days, while only 20% said their view had become more favourable.

(Holcomb 2013)

In relation to the Democratic nomination, it could even be argued that, as the most credible candidate who had opposed a war (in Iraq) which was at that time particularly unpopular with his party's voters, Obama had a distinct advantage over the pro-war Hillary Clinton (Pew Research Center 2013b).

Obama's inability, in the early days of his presidency, to close down the prison camp at Guantanamo Bay was a precursor of many ways in which those enthused by his campaign would become disillusioned by the way in which he actually governed. This again illustrates the problem with a politics of the multitude: it may be mobile, flexible and not based on a rigid dogma, but these characteristics prevent it from converting its ideas into a sustainable political programme. Further, its disdain for building the type of institutions which entrench social change is similarly problematic. This inability of the new social movements to fashion an alternative governmental model which can gain the consent of the majority of activists is well known (White 2012). Similarly, Obama's campaign did not use the obvious popularity of its candidate to challenge the existing national and global political and economic architectures, a lacuna which was revealed in the actuality of his day-to-day governance. For this reason, there was a sharp drop in the number of voters in the 18–24 age range in the 2012 election (File 2013: 7). In contrast, the Californian ideology which was discussed at the beginning of this chapter continues to shape debates about the Internet by unashamedly creating meta-narratives which, however wacky they sometimes appear (as readers of *Wired* might testify), at least have been successful in challenging conventional ideas about politics, society and economics. However, the number of people pursuing this ideology is small in number and drawn from the elites in their respective fields, thus making this particular social movement far removed from the concept of the multitude. But the latter could learn something from the Californian ideology about the importance of the meta-narrative for those seeking to bring about social change; as stated earlier, it is not clear how the postmodern politics practised by the multitude can achieve this. The multitude is, then, adept at causing temporary disruption, but, in the absence of a credible and popular alternative to the present neoliberal arrangements, will remain no more than an irritant rather than a threat to the current global system.

8
Surveillance: The Role of Databases in Contemporary Society

Introduction

However we use digital media, there is one issue that is never far from our minds, even if in practice it mainly operates far from our gaze. That issue is surveillance. Generally, we do not like being watched by strangers and are uneasy about the routine recording of our online activities. However, because we feel largely powerless to protect ourselves against increasingly sophisticated and ubiquitous surveillance technologies, it is something that we try to relegate to the back of our minds, where, unless one is the victim of a mis-placed email or embarrassing photograph that goes viral, it will stay. However, this dormant sense of unease about being surveilled often comes alive during government attempts to legislate further erosions of our privacy, as happened when the UK's New Labour government attempted to introduce ID cards in the first decade of the twenty-first century; ironically, as recent revelations about the activities of the NSA in the USA and GCHQ (General Communications Headquarters) in the UK have shown, the most egregious surveillance programmes have not been subject to even the most basic democratic oversight.

If we are to discuss surveillance in contemporary societies, we need to understand the role of the latest digital technologies in facilitating it. While the electronic database has been in existence for decades, the growth of global computer networks has enabled the bringing together of information which, in the days before they became connected to each other, could only be accessed in individual databases. This means that rather than information being hidden away in separate databases, it is now possible to both widen access to information and allow cross-referencing of it through the interconnected databases in global networks. While many of these databases and the cross-referencing of information in them are visible to those who own them, this global information network is largely invisible to ordinary people. Much information about us is shared between corporations and government departments without our consent, as we are shocked to find out when we receive telephone calls from private companies who seem to know as much about ourselves as we do!

This chapter will therefore begin with a discussion of the development of networked databases and their impact on our lives. It will illustrate how uneasy most people, especially in the so-called Anglophone world, are about invasions of their privacy through the use of information about them, especially when it is gathered by the state. The discussion on state surveillance will bring in Foucault and Deleuze's arguments about the disciplinary and control societies, respectively. No discussion of surveillance would be complete without considering the role of multinational corporations, particularly some of the larger digital media players like Google, and a significant section is devoted to studying corporate surveillance. Partly as a result of the increasing scope of corporate surveillance, the chapter will end with an attempt to convince the reader that there are occasions where the judicious gathering of information on its citizens by the state can demonstrate that surveillance sometimes has a more positive role to play.

The networked database

As the most efficient electronic means of storing and enabling the easy retrieval of enormous amounts of information, inchoate forms of the database were developed in the aftermath of the Second World War. By the late 1960s IBM engineer Ted Codd had created an early version of the relational database, a technology which enabled users both to locate individual records directly, rather than have to search sequentially, and to generate their own tables based on their own variables (Mayer-Schönberger 2009: 75; Paul 2007: 96). Around this time, government departments in the developed world began systematically storing records in electronic form on the basic mainframe computers that were then available. A parallel development in the corporate sector in the 1960s was the huge collation of information needed to service the operation of early credit cards (Athique 2013: 214). The 1970s saw the invention of full-text searching within databases and in that decade and the next, more and more databases were connected to the burgeoning global information network. All the while, the cost of storing information was plunging rapidly and networked computing power was growing exponentially (Mayer-Schönberger 2009: 75–81). Like other social uses of digital media outlined throughout this book, the desire for this type of storage and retrieval function was a long-held one, but the developments in networked computation in recent decades have actualized this greatly in excess of what was hoped for or – in the eyes of those most anxious about these technologies' capacity for surveillance – feared. Before going on to discuss the ways in which these powerful databases are used by governments and corporations around the world, it is useful first to consider some of the conceptual issues that their ubiquity has brought to the surface.

Writing as long ago as 1998, when widespread access to broadband and 3G networks was still years away, Manovich argued that the database and algorithm had migrated from the confines of the closed world of computing to such an extent that the cultural world too operated according to their respective logics (Manovich 2007).[18] (This argument is similar to Castells's assertion in the same decade, cited in Chapter 3, that social life is increasingly operating according to the logic of the global network.) Mark Poster too described databases as cultural forms, rather than merely collators and retrievers of information, around this time; though as shortly will be adumbrated, Poster's understanding of this cultural form diverges from Manovich's in significant ways (Poster 1996; in Aas (2004: 380)).

Manovich (2007) argues that the operating logic of the database is antithetical to narrative, whereas the algorithm positively encourages it. If we leave aside the contentiousness of the latter claim (for is it not the case, as has been argued throughout this book, that the algorithm in its undermining of all structures apart from its own is also antithetical to narrative?) and the observation that large algorithmic search engines are effectively also huge databases, Manovich's argument in relation to the database is a persuasive one. As Aas (2004: 384) demonstrates with the example of parole board decisions, there is a tendency in contemporary society to rely more on easily retrievable data, rather than human judgement. In relation to these types of decisions, human judgement would take into consideration the narrative of an offender's life. The offender's behaviour might have progressed over time such that he/she can no longer be deemed a threat to the community. This judgement might be reinforced by the taking into consideration of contextual information: his/her crimes were committed for specific reasons which no longer exist or have been satisfactorily addressed by the parolee. Increasingly, this narrative of the parolee's life, with its demonstration of a progression in his/her behaviour over time, as well as the professional opinion of parole officers is deemed not as important as information about that person gleaned from the database. As well as being acontextual, information in databases is also often constituted in simplistic binary form:

> Databases create social identities for people on the basis of certain parameters. In terms of crime and security, this may be 'suspicious/not suspicious', 'security risk' or 'potential tax dodger'. In marketing databases this may be a 'high value/low value/non-value' consumer, 'people interested in expensive cars', or 'people with credit problems'.
>
> (Miller 2011: 125)

In other words, there is little nuance in databased information, which is mildly irritating to the consumer in relation to the marketing examples in the quotation above, but potentially extremely problematic when it constitutes the main basis on which more important decisions are made.

For Poster (1995b; cited in Miller 2011: 125), the database is thus not only a new cultural form but a discourse which, in the post-structuralist sense, constitutes the subject. In other words, the mass of information that is held about us in databases not only shapes individual encounters and experiences, but also constitutes our identity. This should be qualified by pointing out that this is not an identity that an individual necessarily would recognize himself/herself to be. Nonetheless, most concepts of identity recognize that the way in which others see us is a significant factor in its construction. This is amplified in relation to the database by the way in which this subject is constituted largely away from our sight, meaning that we are unaware of it. This is particularly problematic where, as in the example of the parolee in the previous paragraph, we are deemed to have undesirable characteristics. I discovered by accident how this works in relation to my own identity when I chanced upon 'my' profile on website ZoomInfo.[19] Though I had nothing to do with constructing this profile, some of the details about my academic career, for example, that I was awarded a doctorate in politics by Queen's University Belfast, are correct. However, in a vivid example of the way in which the database pays no respect to narrative, particularly with regard to timelines, the profile uses the present tense to give information about a post that I left in August 2007: 'Andy White *is* a Research Associate in the Centre for Media Research at Northern Ireland's University of Ulster [my emphasis]' (ZoomInfo 2013). While I still hold my former university in high regard and do not suffer from the type of status-anxiety that many academics do, it is clearly inaccurate to describe an associate professor at the University of Nottingham Ningbo China in this way. But it gets worse. At the bottom of the page two, sentences start with a description of my activities again in the present tense: 'Andy White from our Media Policy Strand has had a new paper published', which were cached (from the University of Ulster's website) in 2009 and 2010, thus giving the impression that I continued working there into 2010 (ZoomInfo 2013). Viewing this information within the context of the University of Ulster's website would lead the user to conclude quickly that I was no longer employed by the institution (my name was removed from the staff list pretty soon after I left in 2007), but once that information is taken out of the context in which it was intended to be viewed then minor oversights become magnified into significant misrepresentations. I would not want to exaggerate the importance of something which causes me little more than mild irritation, but one can imagine how damaging misrepresentations like this might be in other circumstances. A link invites the person to whom the details on the page refer to 'claim your profile' (ZoomInfo 2013). In a sense, this is typical of social networking's subtle pressure on us to create the conditions for our own surveillance by encouraging us to counter any false information about us in the public domain, not by removing it, but by giving away more details about ourselves. While I have not taken up ZoomInfo's invitation, the existence in cyberspace of inaccurate information about my academic career was one of the factors which convinced me that I should create an academia.edu profile for myself.[20]

Foucault's disciplinary society and Deleuze's society of control

Much of Poster's work on discourse is influenced by Michel Foucault, a writer who believed that discourse governs our thinking and behaviour to such an extent that, ultimately, we are not the autonomous subjects that we believe ourselves to be (Danaher, Schirato and Webb 2002: 30–31). In modern societies, individuals are shaped by the discourse, or what we might call norms, of influential institutions, such that these become a 'natural' way of thinking. An example of this is the way in which anyone in the developed world advocating an alternative to the market economy is often seen not merely as wrong, but as dim-witted. In contemporary UK society, the routine ascription of the word 'hero' to British soldiers can make it difficult to enquire more critically into their activities in various warzones (Glenton 2013). The British army's success in persuading the general public that they should view their soldiers in this way is an exemplar of the way in which hegemonic institutions inscribe their norms on wider society through the use of discourse, rather than force. For Foucault, the capacity of sites like the school, hospital, psychiatric unit and barracks to use their institutional heft to inscribe their norms on those that come into contact with them is conceptualized as 'disciplinary', in that people are disciplined by these institutions not to violate those norms. Foucault refers to these institutions as 'carceral', an indication of the way in which the methods of the carceral institution *par excellence* has filtered into wider society, as well as an observation of how similar these institutions, in their compelling of people to inhabit spaces in similarly designed buildings (sites of carceration), are to the prison (Danaher, Schirato and Webb 2002: 53, 108). Foucault argued that the way in which the ideal prison – and by extension all carceral institutions – imposed its norms on its inmates was through a visual regime which, following the philosopher Jeremy Bentham, he terms the panopticon.

As the name suggests, the panopticon is a visual regime which is all-encompassing. Philosopher and prison reformer Jeremy Bentham proposed a carceral institution that would be circular with a watchtower or observation point in the centre. The prisoners would be housed in cells on the building's (circular) perimeter and would be watched by prison officers in the central observation tower. Because the prisoners *could* be watched from the permanently occupied central tower at any time, Bentham argued that they would always behave as if they *were* being permanently watched. In other words, they would tend to feel under pressure to behave according to the norms of the institution, fearing that any transgression would be seen by the prison officers who would then punish them. Foucault took Bentham's idea and extended it to other institutions, like the school, factory or barracks, where people were compulsorily confined for lengthy periods of time. After a while, this regime becomes so effective that people begin to police themselves through self-surveillance: here, one might think of the way in which people use exercise and specialized diets in attempting to conform to an ideal body-type. Thus, Foucault argues that as a means by which norms are enforced, surveillance is an integral part of the modern nation-state (Danaher, Schirato and Webb 2002).

Despite grounding many of his ideas in the growth of the early modern nation-state in Europe around two centuries ago, the relevance of them for the high-powered information-gathering technologies which are now in existence is clear:

Starting at that time [the late eighteenth century], the technology of power of sovereignty began to be supplemented with a new technology of power of governmentality, as Foucault (1991) names it. 'Governmentality' refers to the policing of the population, the institution of surveillance mechanisms, the construction of extensive bureaucracies and social science disciplines to monitor many aspects of the health and welfare of all citizens. A new kind of power was thereby established that constructed each individual as a case. Potentially all individuals would have a dossier filled with information that tracked them through their life course, recording significant events and changes in their circumstances. In other words, *databases* were set in place that identified individuals and traced their comings and goings. Each individual took on an additional identity, one inscribed in the ledgers of government, insurance companies, workplaces, schools, prisons, the military, libraries – every institutionalised space in modern urban environments [my emphasis].

(Poster 2006: 109–110)

While the 'databases' referred to above were primitive compared with their contemporary iterations, the systematic collection of fingerprints and the growth of photograph-based identity documents in the nineteenth century prefigured later, more sophisticated efforts to record the general population (Poster 2006: 110).

Foucault's concept of surveillance is very much based on location, in the sense that surveillance is taking place in institutions or, in the case of record-keeping, linked to a person's home address and/or workplace. Deleuze argues that, rather than requiring sites of incarceration, the power and reach of contemporary technologies means that surveillance takes place everywhere.[21] Further, he argues that the increasing sophistication of databases as a means of gathering information about us means that the disciplinary society's fitting of people into prescribed moulds is no longer applicable. He coined the phrase 'dividuals' to conceptualize the way in which human beings lives are coded numerically, like for example our DNA map:

Dividuals are database constructions, derived from rich, highly textured information on ranges of individuals that can be recombined in endless ways for whatever purposes. They are the abstract digital products of data-mining technologies and search engines and computer profiling, and they are the profiled digital targets of advertising, insurance schemes and opinion polls. A dividual is a data distribution open to precise modulation, stripped down to whatever information construct is required for a specific intervention, task or transaction. Increasingly, postmodern subjectivity is defined by interaction with information meshes and the fractal dividuals they produce. When you use an ATM machine, or access your work environment via your home computer, you are interacting with your dividual self. Likewise, when a database is mined for information on your buying habits, leisure habits, reading habits, communication habits, and so on, you are transformed into a dividual. All aspects of life converted to information, databases, digital samples: this is control society.

(Bogard 2009: 22–23)

There is much debate, fuelled no doubt by Deleuze's unwillingness to refute Foucault's ideas, about whether what the former refers to as the 'control society' has replaced the disciplinary society, amplified it or exists alongside it (Bogard 2009; Poster 2006: 60–61; Savat 2009: 46–47). What can, though, be said with certainty is that surveillance is increasingly reliant on the type of sophisticated database technologies whose importance understandably was not fully recognized by a thinker (Foucault) writing at a time when these were not so well developed. There are two significant ways in which the control society thesis extends Foucault's earlier theory. The control society does not pressurize people into conforming to pre-existing norms; it is more radical than that in that it adjusts its position in response to our behaviour (Savat 2009: 57)! It does this through sorting people into categories; its greater capacity to categorize thus constitutes the first significant way of distinguishing the control society from the disciplinary society (Savat 2009: 53). The computational power of the database means that there can be endless configurations of categorization and thus, by extension, each person has a multiplicity of subjectivities. The database's capacity for producing an endless array of subjects relates to the second fundamental aspect of the control society, its tendency towards the prediction of future behaviour or at least the identification of dividuals who/which are likely to behave in particular ways in the future (Savat 2009: 49). In many respects, the control society is not deemed to be particularly repressive; indeed, it appears to give us more choice. As Chapter 2 outlined, the mining of data from our shopping purchases enables online advertisers to recommend to us products or services which, while occasionally irritating, can be of great benefit to the attention-poor consumer. But control that is more subtle is often more effective, and this type of advertising benefits the large corporation more than it does the consumer (Savat 2009: 57; Stalder 2002: 121). This is especially the case in the health insurance industry, where data-mining is encouraging insurers to become ever bolder in their predictions about individuals' likelihood of being afflicted with certain diseases and ailments, thus enabling them to more effectively tailor premiums to specific individuals, rather than demographic groups; the corollary of this is that they have more information (in terms of health, *more* information usually means the identification of *more* problems) to justify higher premiums or even refuse to insure some people. When these techniques are used to predict behaviour of a criminal nature, then this type of surveillance can become downright insidious. Even in democratic countries, governments expect private corporations with large information-gathering operations to give them access to their data. Consequently, most of us have heard of stories about people whose profiles have been flagged up by security databases after they have bought a number of items that, while innocuous in their singularity, in combination can be deemed indicative of a security risk. As recent revelations about the systematic monitoring of digital communications throughout the world by the USA's NSA have highlighted, it is surveillance by the state rather than by corporations which cause citizens' most concern.

Surveillance and the state

Why is it that people, especially in the so-called Anglophone countries, are more concerned with intrusion by the state than they are with the monitoring activities of corporations? Part of this answer lies in the distrust of both the Right and the Left of the power of the state. Ever since the maturation of nation-states in the nineteenth century, the Left has waged opposition, ranging from calls for reform to outright revolution, to entities which they believe govern in the interests of the bourgeoisie. The Right too is suspicious of state power, believing, in contradistinction to the Left, that its support for society-wide programmes impinges on the rights of free individuals to go about their business without bureaucratic interference; this disposition is accentuated in the USA, where the type of libertarianism manifested by the Californian ideology is more prominent. The first perspective is one which Foucault and Deleuze would have held – and in the former's writing career it was the state which wielded unchallengeable power, especially in the disciplinary institutions upon which Foucault based his ideas – and thus was an appropriate locus for the application of his theory. These days, the Left, especially in democratic countries, tends to be more social democratic, a position that advocates a form of state power which is 'enabling' – in the sense of providing services like health and social benefits like, for example, support for the unemployed – rather than repressive. For this reason, much of the opposition to the state in modern democratic countries in particular emanates from the Right, especially its libertarian strain. However, the residual distrust of the state among the denizens of the Left means that there is almost a consensus across the political spectrum about the inimical nature of government surveillance.

Concern about the growth of state surveillance is in many respects understandable. As we have already seen, the increasing sophistication of storage and retrieval technologies combined with the computerization of government records have given those who rule us an unprecedented amount of information at their fingertips. Added to this is the tendency for governments around the world to request information from private companies. In relation to digital media, many governments monitor Internet Service Providers (ISPs) – the large commercial servers through which we gain access to the Internet – sometimes, as under the USA's 2001 PATRIOT Act, backed up by appropriate legislation; on other occasions, carried out by intelligence services operating with dubious legality (Bennett and Parsons 2013; Chadwick 2006: 277). Recent revelations about the NSA's PRISM programme of electronic surveillance show the systematization of this type of monitoring. The case of email provider Lavabit is indicative of this, though only because of its founder Ladar Levinson's public refusal to hand over information. In a statement posted on Facebook, Levinson claimed that US authorities were exceeding their legislative mandate in their demanding of personal details from Lavabit's user accounts, but nonetheless the pressure exerted him to hand over this information led him to suspend operations (Biggs 2013; Lavabit 2013). Sadly, we do not know for sure whether and to what extent the host of Lavabit's statement, Facebook, or other Internet giants complied with similar requests, but the *Washington Post* reported that it had access to a leaked US government document which listed, on a roster of cooperative companies that are on the NSA's PRISM programme, the following: Microsoft, Yahoo, Google, Facebook, PalTalk, AOL, Skype, YouTube and Apple (Gellman and Poitras 2013). (Attempts to verify the accuracy of this list are complicated by the fact that some of these corporations claim that the Federal government has prohibited them from discussing these types of requests for information (Rushe 2013).)

Many governments around the world require private companies to retain business information for fixed periods of time. Under the 2006 Data Retention Directive, mobile phone companies and internet service providers (ISPs) within the EU must retain data relating to phone calls and emails for anything up to two years (it differs from country to country) after the date of the original communication. Miller (2011: 118) reports that in Russia, communication companies must retain data for three years; in the USA the retention time is two years. Therefore, even if companies in those respective countries want to protect their customers' privacy by deleting details about them, they are prohibited from doing so until the legal retention period has elapsed. There is thus a veritable goldmine of information in the hands of private corporations that government can, subject to safeguards which are often loose or easily circumvented, access. More alarming is the fact that individuals have very little recourse to the law if this data is misused. In the USA, individual privacy is not a legal right. Indeed, the commonly held belief in the USA that freedom of expression over-rides virtually every other right (a manifestation of libertarianism) means that there are very few restrictions, even on the dissemination of personal data in commercial databases (Miller 2011: 115–116). There is much more protection for individual privacy in European states, but even that can be undermined by information-sharing agreements with governments or corporations in states with laxer privacy regimes. The handing over of EU citizens' flight details to the USA and the complexity of identifying the proper legal jurisdiction for transnational information flows like cloud computing are just two of the many issues in recent years which have left European citizens confused about their legal rights (Bennett and Parsons 2013: 498).

When people are consulted about increased state surveillance, they tend to baulk at it. First mooted in 2002, the UK's New Labour government's ill-fated and long-running national identity card scheme was finally killed off by the incoming coalition government in 2010 (BBC News 2010). Home Secretary Theresa May's justification for the axing of the scheme was couched in libertarian terms:

> Ms May said: 'This bill is a first step of many that this government is taking to *reduce the control of the state over decent, law-abiding people and hand power back to them*. With swift Parliamentary approval, we aim to consign identity cards and the *intrusive ID card scheme* to history within 100 days' [my emphasis].
>
> (BBC News 2010)

A similar tale of the rolling back of state surveillance through democratic politics seems to have been played out in the USA in relation to the TIA (Total Information Awareness) programme that was mooted in the immediate aftermath of the 9/11 attacks in 2001. Proposed ironically by the same branch of government that was responsible for developing the Internet, DARPA (Defense Advanced Projects Agency), the TIA would have given the federal government virtually unrestricted powers to collect and mine information on American citizens from commercial and official databases. Opposition to it from civil libertarians resulted in the US Congress refusing to devote any funds to the programme, which was forced to close down (Weinberger, S. 2007). However, the formal closure of this programme did not mean that the federal government shelved its plans to bolster its surveillance capacity, as the revelations in 2013 about the long-running PRISM programme illustrated. Instead, the TIA programme was moved to the NSA, where it received a budget deliberately kept secret from the general public (Hayes 2013). As Hayes (2013) argues, the revelations about PRISM are of great concern, not only because of the widespread trawling for information, but also because of the way in which this hugely expensive and highly political programme has been developed without democratic oversight. And what of Theresa May, the Home Secretary who axed the ID card scheme for being 'intrusive'? Former Shadow Home Secretary and fellow party MP David Davis finds it hard to believe that either she or the Foreign Secretary William Hague would not have been involved in the authorization of the surveillance of UK citizens that the PRISM programme undertook (*The Huffington Post UK* 2013); indeed, *The Guardian* newspaper reported that the UK's GCHQ had been gathering information through this programme from as long ago as June 2010, around the time that May was denouncing the ID card scheme (Hopkins 2013). Thus, as in the USA, it seems that democratic discussion of new surveillance legislation in the UK is merely a veneer which occludes a parallel, more secretive and important, process of accretion of the state's power to monitor its citizens. It would be easy to fall into the trap of justifying this by referencing the heightened state of security in the ongoing so-called War on Terror, where, from the US and UK perspective, the enemy is largely unseen. However, that would be to ignore the fact that this secretive policy-making process has long been a feature of approaches to security in the UK and the USA. For example, an earlier iteration of the aforementioned programmes was ECHELON, a surveillance system that was developed from the 1970s onwards by the NSA and the UK's GCHQ, and which involved many other countries as partners. The existence of this long-running programme was not even acknowledged until 1999, when the Australian government confirmed its involvement, and certainly never faced parliamentary scrutiny (Bomford 1999; Chadwick 2006: 282).[22]

Another reason for distrusting government surveillance programmes is the commonly held concern that databases are not really secure. We have already seen how information can be passed on to third parties without our knowledge and that claims about us can be made based on information which might be viewed out of context. But there have been numerous, more serious cases in recent years of official databases being hacked into, of officials being tricked into handing over information that is on government databases and the loss of flash drives and laptops with confidential information. The following table taken from an article in *The Guardian* newspaper detailing breaches of official databases in the UK in the year leading up to October 2008 as reported to the Information Commissioner, Richard Thomas, shows the scale of the problem (Table 8.1).

At a time when opposition to the introduction of ID cards in the UK was particularly fierce, it was ironic that hardly a week seemed to pass without reports of some sort of security breach of government data.[23] Added to this are the many instances of individuals and even organizations obtaining this type of information by illicit means. The 2012 Leveson Inquiry's detailing of the intercepting of mobile phone voicemails and the hacking into email accounts by sections of the UK's newspaper industry highlighted practices that have been employed for a long time and in a wide range of operations (Leveson 2012). A few years prior to the discovery of the News International hacking which triggered the Leveson Inquiry, Nick Davies detailed some of the ways in which his fellow journalists in the UK find confidential information. Impersonation is a means by which some people have managed to gather personal details from the databases of banks, credit card companies and British Telecom. Sometimes, journalists have resorted to bribing the police officers and civil servants who have access to government and police records (Davies 2008: 273–275). The releasing of diplomatic cables by Wikileaks and the dissemination of classified NSA documents by Edward Snowden are further, more recent examples of the insecurity of government data, even if the latter case is complicated by the fact that the whistleblower was motivated to expose what he and others believe is a government surveillance regime that is out of control.

Bottom-up surveillance

The cases in the last section exemplify the coming to prominence in the last few years of a more 'bottom-up' form of surveillance. In its most sanguine guise, it enables citizens' to hold governments to account through the use of surveillance techniques which the proliferation of

Table 8.1 Data security breaches in the UK in the year up to October 2008

Who lost it?	Email error	Postal error	Lost/stolen computer	Inappropriate disclosure	Website security	Lost paper records	Lost disk etc.	Other	Total breaches
Local government		2	4	5	2	3	4	6	26
Central government		7	7	0	1	5	6	2	28
NHS & healthcare	1	2	27	5	1	14	18	7	75
Private sector	6	7	18	13	6	4	17	9	80
Charities	1	3	4			4	4	4	20
Other public sector		5	7	11	4	4	13	3	47
Other		1						1	1
Total	8	27	67	34	14	34	62	31	277

Source: Travis (2008) [my emphasis].

digital media has made widely available. Ian Tomlinson's death at the G20 protest in 2009, detailed in the previous chapter, was one such example of the demos using digital media to challenge the police's view of events on that fateful day. While the particular application of digital media in that case did not secure the conviction of the police officer who shoved Ian Tomlinson, this type of surveillance has been more influential elsewhere. The 1991 Gulf War was managed so rigidly by the American-led coalition that there was little opportunity for the media to report anything that challenged the narrative of the victors. Powerful nations have always had the capacity to 'survey' the 'enemy' in warzones and send those images back home. In the 1991 Gulf War, television viewers in North America and Europe were bombarded with many images of operations against Iraqi targets, crucially without evidence of much bloodshed. The success of this media strategy led Jean Baudrillard (1995) famously to declare that 'the Gulf War did not take place', meaning that what we saw on our TV screens was a bloodless event, almost like a video game, while the great suffering of the thousands of Iraqis who died was televisually ignored. Mainstream accounts of surveillance, including those in this chapter, highlight the discrepancy in the power between the state and the individual in relation to surveillance, and the Gulf War exemplifies this. However, events during the 2003 Iraq War reflect the narrowing of this discrepancy in the last decade. In April 2004, digital images of what appeared to be the systematic abuse and torture of Iraqi prisoners by American guards at Baghdad's Abu Ghraib prison began to be circulated to the world's media (Webster 2006: 217). In a further twist, the use by Iraqi combatants of hostage videos and other disturbing Internet imagery to show graphically the death and suffering of the soldiers of America and her allies brought the ugly reality of war right into the homes of those nations' citizens in a way that has never been witnessed before (Hill, A. 2009: 62). While one should be wary of trying to identify cause and effect in a conflict as violent as the one in Iraq in the middle of the previous decade, the almost daily procession of these and other images might well have undermined support for the war in the USA and among its allied partners, as well as increasing anger against the occupying army in Iraq and beyond to the point that it fuelled the resistance to it. An arguably positive aspect of this shifting visual regime is that it sends a clear message to western armies (and armies from other regions too) that they cannot act with impunity in warzones around the world.[24]

It is not only in warzones that this new form of surveillance is effective. If we think back to the discussion on crowd sourcing, it is possible to relate that particular method to bottom-up surveillance. *The Guardian* newspaper employed crowd sourcing as a means of investigating British MPs' expenses. Covering a four-year period, the expenses comprised 700,000 receipts for all 646 MPs (Rogers 2009a). The newspaper designed an app to make it easier to crowd source this material and encouraged readers click a button marked 'investigate this!' if they thought a particular find warranted further attention from the newspaper's reporters (Rogers 2009b). This massive crowd sourcing operation was able, like a powerful computer database, to sort information into categories, as did one report on MPs' expense claims from April 2008 to June 2009 relating to their second homes (*The Guardian* 2009). The idea that this represents a new form of progressive politics has been challenged elsewhere by the author (White 2012), but *The Guardian*'s crowd sourcing exercise, alongside the *Daily Telegraph*'s original investigation about individual MPs' expenses (copied from a classified computer at Westminster), undoubtedly made even those MPs' who were largely blameless uncomfortable with the scrutiny of seemingly all aspects of their lives (Brook and Gillan 2009; Darling 2011: 237).

In most cases, though, it is vulnerable individuals, rather than powerful politicians and governments that are the subject of bottom-up surveillance, as evidenced in the increasing use of software that enables the monitoring of a third party's Internet use and applications which allow mobile phone users to be located. While these types of software can be put to good use in allowing parents to monitor their children, they can be also be used for more controversial purposes too, like when employers monitor their employees' email (Miller 2011: 119). One does not have to spend too much time on the World Wide Web before stumbling across sites dedicated to exposing different kinds of behaviour. While many of these websites might be well meaning in intention, their presentation of information, images and videos taken out of context or just plainly erroneous can cause much distress. One such website, DontDateHimGirl.com, invited women to share information about men who were suspect of character, if not downright abusive in some way. While this was in many ways a commendable project and useful resource, there was no way of verifying whether allegations about the men identified on the site were actually true. This manifested itself in cases like the posting of very damaging information about one person by a friend as a 'joke' and the bringing of a lawsuit by another man who claimed that information about himself on the website was false (Green 2006). This led to the website announcing in July 2010 that 'it will remove its database of alleged cads from the site in August' (*EON: Enhanced Online News 2010*). But websites of a much more malicious nature still exist, such as those 'revenge porn' ones which invite users to upload pornographic photographs of their ex-partners (Filipovic 2013). Andrejevic uses the term 'lateral surveillance' to describe these phenomena and, while we should not necessarily dismiss the opportunity they give us for bringing the powerful to account, we should heed his words about the more negative aspects of this type of surveillance too:

> The participatory injunction of the interactive revolution extends monitoring techniques from the cloistered offices of the Pentagon to the everyday spaces of homes and offices, from law enforcement and espionage to dating, parenting, and social life. In an era in which everyone is to be considered potentially suspect, we are invited to become spies – for our own good.
>
> (Andrejevic 2006: 406)

Corporate surveillance

What has been alluded to throughout this chapter is the increasingly central role of the corporation in the practising of surveillance. Chapter 2 cited Jaron Lanier and Andrew Keen's concerns that social networking was providing free platforms on the condition that users' details can be harvested by commercial operators. But, as that chapter also highlighted, the desire to authenticate our own identities and others too means that this type of surveillance does not appear to be taken as seriously by users as state surveillance. It also reflects the consensus of the Right and the Left about the repressive nature of the modern state. While it is difficult to make people enthusiastic about the threat posed by targeted advertising, the earlier mentioned revelations of purported collusion between some of the social networking giants and the NSA has made people more wary about not only those websites, but cloud computing generally (Risen 2013). Surveillance techniques that originated in the private sector for commercial purposes have been adapted by other parts of the corporate world to monitor employees, and even by autocratic governments who want to spy on their citizens (Curran 2012: 44). In general, the increasing privatization and sub-contracting of governmental functions to private organizations means that the rigid distinction that used to exist between surveillance by statutory bodies and corporations is hard to discern these days. This makes it all the more important to assess the surveillance activities of the private sector.

As suggested earlier, the information-gathering operations of private companies are largely unregulated. In the USA in particular, information circulation is viewed as an issue of freedom of speech and thus is not subject to many restrictions (Miller 2011: 115–116). And there is a commercial imperative to trade information, with figures from as long ago as 1996 suggesting that $3 billion a year is generated in the USA from the sale of mailing lists alone (Fenrich 1996: 956; cited in Solove 2004: 19). Just think of how dependent we are on credit agencies when we want to open a bank account or take out a loan. But how is this sector regulated? A better way of asking this question would be to ask what recourse does an individual have if information about them in the database is incorrect. The short answer appears to be 'very little', as then CEO and president of Regulatory DataCorp (RDC), one of the leading credit checking companies, stated: 'There are no guarantees. Is the public information wrong? We don't have enough information to say it's wrong' (Hamilton 2003; cited in Solove 2004: 21). Despite this kind of function being undeniably a part of the public realm, rather than merely private decision-making, investigations into these companies' activities is often thwarted by their citing of commercial confidentiality. This is even the case when corporations take over governmental functions, with many freedom of information acts preventing citizens, on the grounds of commercial confidentiality, from scrutinizing certain aspects of public policy (see Ministry of Justice (UK) 2012).

This lack of scrutiny becomes even more of a concern when corporations carry out their own surveillance activities. In addition to some of the tracking and monitoring software mentioned earlier, technologies that track goods are now being used to monitor human beings, as with the use of RFID (radio frequency identification) microchips to monitor attendance in US schools (Chadwick 2006: 287). And we have already seen how illegal forms of surveillance were carried out by some of the UK's leading newspapers. But, as Lanier and Keen argued in Chapter 2, it seems that it is the social networking and search engine giants which are surveillance regimes *par excellence*. In order to subject this claim to greater scrutiny, it would be useful to close this section with a detailed look at the activities of one of the largest of them, Google.

The power of the world's most popular search engine is such that, in the last few years, it has begun to receive criticism of its surveillance capabilities from writers whose main interest is in state surveillance (Porter 2009). This is partly a result of an expansion of Google's activities beyond search and retrieval. Making information about us easy to collate and retrieve can at times be disconcerting, especially where that information is false or embarrassing, but it is something that most of us have learned to live with. Indeed, while the harvesting of information for commercial reasons is criticized by many scholars, the popularity not only of Google but of social networking platforms as well suggests that the general public does not find this overly intrusive. However, were information about our searching activities to be made public, then that would be considered a significant violation of our privacy. While it must be stressed that Google has not been involved in such a breach of privacy, in August 2006 AOL accidentally released the search records of 658,000 of its users. Even though the records were anonymized, it was not difficult to identify the users who effectively had their most private thoughts posted in the public domain (Keen 2007: 164–168). As the many breaches of government databases demonstrated earlier, none of these information stores is totally secure. One way of mitigating the potential damage a similar leak might cause in the future would be to delete information after a short period of time. What, then, is Google's data retention policy? This is a question that is very difficult to answer, even with the aid of its eponymous search engine! It is clear that the European Union thinks that Google retains data for far too long, but how long is not known (Garside 2013a). A new privacy policy that came into effect on 1 March 2012 is extremely opaque, but does not appear to commit Google to expunge any search data at all (Galperin 2012; Garside 2012).

Google's launch of Street View in 2007 was a significant extension of its capabilities, to the extent that Vaidhyanathan (2011: 98–107) has encapsulated what it ushers in as the 'universalization of surveillance'. Linked to its Google Maps programme, street cars photograph the entire area of the countries in which the Street View operates. While this is no doubt entertaining to those who like looking at other people's properties and can be a means by which one can navigate around a city without being physically present in it, it is considered such a violation of privacy that even in countries where the authorities have allowed Google's cars to film there has been localized resistance (Vaidhyanathan 2011: 101–105). Inevitably, there have been cases of the identification of buildings, like refuges for women fleeing from abusive relationships, which could cause harm to vulnerable people (Peev 2010). More worryingly, despite some attempt at blurring their features, it has been possible to identify people caught on camera. Even when their images are subsequently removed or rendered opaque, it has not entirely removed them from the wider Internet (Vaidhyanathan 2011: 104). The particularly tragic case of the father who in 2013 discovered a 2009 image of his dead son lying near a railway line between North Richmond and San Pablo, California on Google Maps illustrates the inherent potential for distress that this type of visual regime can cause (*The Guardian* 2013).

Some positive and concluding thoughts on the role of the state in relation to surveillance

This whole chapter appears to have been based on the premise that to be surveilled is unconditionally a bad thing. One does not have subscribe to the popular refrain that 'if you've nothing to hide then you've nothing to fear' to appreciate that surveillance is not always necessarily inimical to the individual. This is especially the case with state surveillance. The last statement appears to be counter-intuitive in the face of everything that has been written about in this chapter. One way of thinking about this is to imagine a society without surveillance. We might think that the amount of personal information gathered by credit card companies is excessive, but are we prepared to forgo this form of payment? This example illustrates the fact that if we want to live in well-organized modern societies, then we cannot avoid sophisticated surveillance apparatuses:

> Bluntly, *routine surveillance* is a prerequisite of effective social organisation. Not surprisingly, therefore, it is easy to trace the expansion of ways of observing people (from the census to the checkout tills, from medical records to telephone accounts, from bank statements to school records) moving in tandem with the increased organisation which is so much a feature of life today. *Organisation and observation are conjoined twins, ones that have grown together with the development of the modern world* [emphasis in the original].
>
> (Webster 2006: 205–206)

Indeed, this is crucial for 'surveillance as social care' (Chadwick 2006: 258) or surveillance as 'categorical care' (Webster 2006: 222) in the service of social welfare, health and education systems. An example of this form of positive surveillance is the UK National Health Service's Summary Care Records database, whose storage of the medical details of the country's citizens could be invaluable to those facing medical emergencies:

> A Summary Care Record is an electronic record which contains information about the medicines you take, allergies you suffer from and any bad reactions to medicines you have had.

> Having this information stored in one place makes it easier for healthcare staff to treat you in an emergency, or when your GP practice is closed.
>
> (NHS 2013)

Think of the potential problems that might arise from the state not having access to these details, especially if emergency medical workers fear administering drugs because of the possibility that it might induce allergic reactions.

A more positive approach to state surveillance would also help to counteract some of the more egregious aspects of corporate surveillance. Problems relating to excessive retention of data and the recycling of information in corporate databases could be ameliorated through more stringent regulation, as the European Union has been trying to do with regard to Google's privacy policy (Thompson 2013). This, though, will be a tough task, particularly in Anglophone countries. Despite the damning evidence of electronic eavesdropping by sections of the UK's newspaper industry detailed in the Leveson Report, there is still a tendency for the nation's legislators to view any form of media regulation as state censorship. The faultiness of these types of libertarian arguments is illustrated by the case of the so-called 'death panels' that Barack Obama's healthcare reforms are said to have created. The opponents of Obama's reforms argue that these statutory panels will make decisions about whether or not to fund medical treatment, implying that there will be cases where the 'bureaucracy' will prevent life-saving operations from going ahead. But, as *Salon* reported, private insurance companies have been doing just this for years (Madden 2009); the increasing sophistication of the databases to which they have access will surely lead to more of these cases in the coming years. It is not surprising that in a libertarian culture like the USA, the state will be lambasted for decisions or errors that, when carried out in the private sector, will bring about no such opprobrium. But that does not mean that this approach is not fundamentally wrong. A way around this problem might be to acknowledge that surveillance and encroachments on our privacy are impossible to combat, especially with the powerful digital technologies that are available today (Stalder 2002). The long campaign against ID cards in the UK is a potent illustration of the ineffectiveness of approaches based solely on protecting people's privacy: the political campaign could be considered successful, but in not addressing corporate surveillance or the type of surveillance that is carried out by the security services it was clearly a strategic failure. An acknowledgement that not only is surveillance inevitable but in some cases desirable, would enable the development of strategies which focus more on holding the information-gatherers to account. And the responsibility for this accountability must come from individual nation-states or collectives like the European Union, rather than present regimes which emphasize corporate self-regulation. In the libertarian climates that prevail in many parts of the world, it is a tough argument to make, but President Obama's proposals in early 2014 to hand over NSA-generated metadata to a third party and the inclusion of privacy rights advocates on the secret Foreign Intelligence Surveillance Court (FISC) suggest that it is gaining traction (*BBC News* 2014).

9
Digital Media Use in the Developing World

Introduction

The bulk of the content of this book relates to the Anglophone world in which the author was educated and works.[25] In this sense, the book is similar to most studies of digital media, especially monographs, but the omission of a serious discussion of how many of the concepts developed throughout these pages might be applied beyond the Anglophone world, especially in developing countries, would be deeply problematic. Indeed, such a discussion is useful, not only in identifying some of the interesting uses of digital media in different parts of the world, but also in adding heavy caveats to claims that we can invent technologies and concepts in the developed world and simply apply them unconditionally elsewhere; as this chapter will go on to demonstrate, there are also examples of ideas and innovations travelling in the opposite direction. Space does not permit a discussion of the development and use of digital media in all parts of the developing world. Instead, the chapter will centre on the so-called digital divide between the developing and developed world, discuss the extent to which this concept is still relevant in today's global media environment, as well as highlighting how the extent to which these technologies are *used* problematizes the binarism inherent in theories of the digital divide. After presenting a range of examples of digital media use throughout the developing world, the chapter will end with a discussion of *Ushahidi*, an example of a successful export of a digital media technology from the *developing* to *developed* world.

Orientalism, cultural imperialism and the mass media

One of the main concerns of communication scholars in previous decades was that of the dominance of the values of the developed world in global media. As the example in the last chapter of the 1991 Gulf War demonstrated, until relatively recently the developed world tended to 'fix its gaze' on the developing world through the use of its powerful media corporations. In his landmark study in 1978, Edward Said argued that this is a manifestation of 'orientalism', the tendency, going right back to the nineteenth century, of countries in the developed world to see their own cultures as superior to those in the developing world. This mind-set both exoticizes and victimizes people in the developing world, even within serious literature and scholarship (Said 2003). A contemporary example of residual cultural imperialism can be found when one enters the term 'African people' into www.google.co.uk and is presented with a bank of images which are overwhelmingly of Africans in very elaborate tribal clothes. The sub-categories at the top of the page also appear to validate Said's concerns: 'poor', 'culture', 'dancing', 'happy African people', 'African people working' (they all are working in fields rather than in offices) and 'starving'.[26] The ubiquity of these attitudes and the traditionally powerful position of the developed world, especially in the colonial era, means that even many 'oriental' intellectuals internalize the developed world's view of their own societies and people:

> Every colonized people – in other words, people in whom an inferiority complex has taken root, whose local cultural originality has been committed to the grave – position themselves in relation to the civilizing language: i.e., the metropolitan culture. The more the colonized has assimilated the cultural values of the metropolis, the more he will have escaped the bush. The more he rejects his blackness and the bush, the whiter he will become.
>
> (Fanon 1952/2008: 2–3)

This problem was recognized by many in the developing world in the 1970s, from where there were calls for a new world information and communication order (NWICO). The debate that this ignited culminated in the 1980 MacBride Report, which was essentially UNESCO's policy response to what was seen as cultural imperialism in all its manifestations. Like the authors above, the report is very much concerned with the way in which the supposed free global flow of information in fact privileges the values and interests of the developed world. Reflecting a widely held disposition that it was virtually impossible drastically to alter what it perceived as imbalanced global media flows, the report argued that countries, especially in the developing world, should adopt protectionist policies with regard to their own media: 'Every country should develop its communication patterns in accordance with its own conditions, needs and traditions, thus strengthening its integrity, independence and self-reliance' (International Commission for the Study of Communication Problems 1980: 254). Another recommendation tackles cultural imperialism head-on, calling for countries to develop their own 'national cultural policies' partly 'to modify situations in many developed and developing countries which *suffer from cultural dominance*' (my emphasis) (International Commission for the Study of Communication Problems 1980: 259). While these provisions were not likely to be popular in the developed world, the report's commitment to freedom of speech, reinforced by an explicit disavowal of any form of censorship, should have appealed to western democracies (International Commission for the Study of Communication Problems 1980: 265–266). Despite this, the USA, the UK and Singapore decided to withdraw financial support from UNESCO in 1984 (Flew 2007: 202).

The report was effectively side-lined, as were discussions on NWICO (endorsed by MacBride), as those countries which benefited from the imbalance in global media flows ignored it. But the report's concern about the way in which the dominance of the world's media by a small number of large corporations from the developed world affected the reporting of important issues in the developing world was still relevant even in the early days of the World Wide Web. The example of the 1991 Gulf War has already been mentioned and CNN's unchallenged role in projecting images around the world during that conflict illustrates the partiality that stemmed from this global imbalance in the mass media environment. This is important because in a political environment where some argue that images on CNN can determine foreign policy interventions by the USA, the relaying of partial information or its omission altogether can have a disproportionately powerful effect (Livingston 1997; Lull 2000: 194–195).

While the cultural imperialist thesis was never uncontested, the move towards a more plural global media environment has seriously challenged its credibility. Tomlinson (2003; cited in Flew 2007: 120) argues that, in addition to thinking about globalization as the means by which large media corporations can spread their ideas, it can also promote other identities not linked to western capitalism. Indeed, Thussu (2007:

3) has highlighted the global rise of Bollywood, Chinese film, Latin American TV and Arab satellite TV as examples of the way in which some of the more powerful developing nations can use globalization to their advantage in projecting their national cultures beyond their respective regions. In a review of contemporary debates on cultural imperialism, Flew (2007: 124–128) argues that there is a tendency for its proponents to deploy the simplistic 'media effects' model criticized in other parts of this book; more often than not people view any cultural content with a discerning eye. As part of this judicious form of viewing, people often prefer locally made programmes that better reflect their own culture than glossily presented and expensively produced foreign imports. Also, despite the best efforts in opening up global markets to media products, many countries, including the one in which I am presently writing (China), deploy the kind of protectionist policies advocated by the MacBride Report. But, notwithstanding the earlier example from Google Images, it is the rapid global proliferation of digital media, particularly in the form of the World Wide Web, that appears to have fatally undermined the cultural imperialist thesis. While earlier chapters have cautioned against accepting at face value some of the more celebratory accounts of the plurality of digital media content, it is nonetheless the case that, as the myriad of blogs, eye-witness accounts (Lovink 2011: 115–121) and, *in extremis*, footage of military assaults and western hostages illustrated in the 2003 Iraq War, most of the concerns in the MacBride Report about the wide disparities between the developed and developing worlds in their respective capacity to produce their own media content have been largely resolved.

The digital divide

While debates about cultural imperialism have largely receded into the background, another concern has moved into the foreground. While his conception of it is certainly open to critique, it is difficult to disagree with Manual Castells's (2010a) argument that we live in a global network society from which politics, economics and social practices around the world cannot disentangle themselves. Castells draws a distinction between those who are fully imbricated in the network (the privileged) and those who are not (the marginalized). This echoes the views of policy-makers too: in 1995 the US Department of Commerce coined the phrase the 'digital divide' to describe the gap between the digital haves and have-nots (Maier-Rabler 2006: 198). While the digital divide is often used in the USA to describe internal divisions, measures to close the gap at the global level quickly became the priority for activists and policy-makers interested in the role of information and communication technologies (ICT)[27] in social and economic development in the developing world. In one respect, this debate reflects the discussion about cultural imperialism in the sense that, notwithstanding the divisions within countries, most activists employ a heuristic which states that it is the developed world which designs, owns, sells and uses most of the digital infrastructure and software, while the developing world is lacking in infrastructure, has much fewer users per head of population and often has to pay American prices for software and hardware. But in another sense, insofar as it exists, it is a much more serious problem than cultural imperialism. This is because it is not solely about how the developing world is 'represented' in the media – as argued in earlier chapters, emphases on representation tend to elide the way in which audiences choose and watch content in a more discerning way than a passive, representational model of media consumption assumes. Digital media's imbrication in virtually all social, economic and political practices means that to be deprived of access to this technology is much more problematic than concerns about cultural imperialism were (Chadwick 2006: 50). And as the discussions about the global intellectual property regimes later in this chapter will illustrate, it is not as easy to opt-out of the global information network as it would have been to ignore the international media in the pre-Internet era. Furthermore, being unable to access the newest communication technologies is incompatible with Article 19 of the UN's Universal Declaration of Human Rights:

> Everyone has the right to freedom of opinion and expression; this right includes freedom to hold opinions without interference and to seek, receive and impart information and ideas through any media and regardless of frontiers.
>
> (United Nations 1948)

Indeed, it could be argued that there is a greater need for information on health issues and climactic conditions in the developing world, as well the need to respond quickly to the natural and social crises that are more common there, all of which is hampered by poor access to digital technologies (Miller 2011: 103).

But to what extent is the digital divide a problem? Figures from the ITU (International Telecommunication Union) show that, despite improvements in recent years, the gulf between the developed and

Table 9.1   2013 International Telecommunication Union (ITU) figures for ICT capacity

	Developed world	Developing world	Africa
Active mobile-broadband subscriptions per 100 inhabitants	74.8	19.8	10.9
Fixed (wired) broadband subscriptions per 100 inhabitants	27.2	6.1	0.3
Households with Internet access per 100 inhabitants	77.7	28.0	6.7

Source: International Telecommunication Union (ITU) 2013: 3–8.

developing worlds is still great. In the statistics above, one should bear in mind that China is classified as a developing nation and its rapidly growing number of digital media users in what is the world's most populous state can thus inflate the developing world's figures. For this reason, the figures for Africa alone (the continent that faces the most difficulties with regard to digital connectivity) have been included to illustrate the scale of the problem (Table 9.1).

While the figures above make depressing reading, Castells's (2010a: xxv) observation that 'the ratio between Internet access in OECD and developing countries fell from 80.6:1 in 1997 to 5.8:1 in 2007' is an indication of some progress over the past decade-and-a-half. And considerable efforts have been made in the past few years to improve the digital infrastructure, particularly in the rural hinterlands, as this section will go on to outline.

Ironically, one of the biggest single factors in the provision of cheap Internet access for many parts of the developing world was the turn-of-the-millennium dotcom crash in the USA. In the boom that preceded the crash, over-investment in US cable companies led them to lay massive amounts of submarine and land-based fibre-optic cables all around the world. One of the relationships that this strengthened was that between US companies and relatively low-paid Indian IT engineers who at that time were working together on fixing the Y2K bug which, it was feared, would cripple computers worldwide on 1 January 2000 (Friedman 2006: 131).[28] While computers themselves did not crash *en masse* at the beginning of the millennium, the market in Internet stocks did. The over-supply of fibre-optic cables resulted in a fire sale, from which countries like India and, to a lesser extent, China and the former

Soviet *bloc* benefited (Friedman 2006: 126). While the continent of Africa did not benefit to anything like the same extent, the last decade has seen a steady increase in the number of fibre-optic cables servicing its telecommunications infrastructure, with the operationalizing in 2009 of a number of submarine cables which gave east Africa similar access to the Internet that other parts of the continent already benefited from (ITU 2013: 47; Unwin 2013: 544).

In advance of the building of a comprehensive infrastructure of submarine fibre-optic cables, Africa had promoted a 'leapfrogging' form of development. Rather than putting all their efforts into building a fixed-line telecommunications infrastructure which in the developed world had taken decades to build, and at very great expense, in Africa and other parts of the developing world there has been a tendency to 'leapfrog' that phase of development for the next: mobile telecommunications (Miller 2011: 107).[29] As Miller (2011: 104) points out, discussion of the digital divide is impoverished if it focuses on access to the Internet to the exclusion of mobile phone use. While submarine fibre-optic cables provide coastal regions with great Internet access, in the African hinterlands it is easier and cheaper to build a mobile infrastructure. Costs are further cut through the widespread practice of sharing phones and using intermediaries (Miller 2011: 105–106). Mobile phones are used in financial transactions not only in Africa, but also in countries like the Philippines, China and India both for business and personal reasons (Miller 2011: 65, 108; Unwin 2013: 543). Mobile phones can often give people access to the Internet through wireless technology, but even where this is not the case, there are applications, like *Ushahidi* (which will be discussed in greater detail later in the chapter), which skilfully utilize texting to get messages out to the wider community (Unwin 2013: 546). The following observation from Kenya in 2007, a couple of years before the submarine fibre-optic cables were laid in east Africa, illustrates the impressive range of opportunities that mobile phones afford (Mason 2012: 133–134):

When I travelled across Kenya in 2007, following the cellphone signal from Mombasa into the Rift Valley, it was clear that mobile telephony was causing a micro-level social upheaval. I met minibus drivers suddenly able to contact their bosses when pulled over by corrupt police in search of bribes; hairdressers who, by simply collecting the cellphone numbers of their client, had freed themselves from the decades-old tyranny of the 'madam' who owns the parlour; slum dwellers mobilizing by text message to fight evictions; villagers able to receive cash remittances at the touch of a button through a cellphone money-transfer system.

Even in the red dust of Masai country, tribespeople living in mud and grass huts had been able to procure cheap Chinese cellphones, which they charged using solar power.

It is argued that one of the advantages of this type of connectivity is that it is relatively cheap in regions where access to broadband Internet is still expensive (Miller 2011: 106). The figures below offer partial support to this argument (Table 9.2).

These figures show that while mobile access to broadband Internet is cheaper than fixed-line access in Africa, at a cost of at least 36.2 per cent of gross national income per capita, even this is not financially viable for the vast majority of Africans. In contrast, in most African countries mobile phone ownership exceeds 60 subscriptions per 100 inhabitants, even if only eight countries exceed 100 subscriptions per 100 inhabitants (ITU 2013: 56, 59). However, growth between 2011 and 2012 was mainly concentrated in the richest countries, thus widening regional differences; indeed, the country with the world's lowest proportion of mobile phone subscribers, Eritrea, showed virtually no growth from its level of 5 per cent in the year 2011–2012 (ITU 2013: 56). This is

Table 9.2　2013 International Telecommunication Union (ITU) figures for the cost of access to broadband Internet

	Developed world	Developing world	Africa
Fixed-broadband prices as a percentage of GNI (gross national income) per capita	1.7	31.0	64.3
Mobile-broadband prices as a percentage of GNI per capita			
Postpaid handset-based (500MB)	1.3	11.3	36.2
Prepaid handset-based (500MB)	1.1	15.7	38.8
Postpaid computer-based (1 GB)	1.5	18.8	54.6
Prepaid computer-based (1 GB)	2.3	24.7	58.3

Source: International Telecommunication Union (ITU) 2013: 77–99.

reflected in the fact that growth in mobile-broadband subscriptions has slowed in the developing world since 2010–2011, while a tiny rise in the growth rate in the year to 2012–2013 in fixed (wired)-broadband subscriptions was preceded by slowing growth every year from 2003–2004 (ITU 2013: 2, 4). The growth in the number of households in the developing world with Internet access has also slowed since 2010–2011 (ITU 2013: 8). Even these figures are inflated by the continued steady growth of the world's most populous nation, China, which now has 590.6 million Internet users, representing 44.1 per cent of its total population (Pew Research Center 2013a). As most commentators on the global digital divide point out, these huge differences between the developed and developing worlds, and within regions of the latter, largely reflect different levels of economic development. This might explain why, despite there being initial rapid growth from a low base, in the poorest parts of the developing world it will not be possible to make a serious attempt to bridge the digital divide until median incomes are much higher than they are now.

Global Internet governance and the digital divide

While the contemporary discussion about the digital divide has seemingly eclipsed 1980s debates about cultural imperialism, the spreading of communication technologies throughout the world since then has not been accompanied by a commensurate diffusion in global governance. As we have seen in earlier chapters, while the Internet was developed in an *ad hoc*, free-wheeling way, it was very much controlled by institutions and government departments in the United States. Mindful of the criticisms of other nations, the US government set up ICANN (Internet Corporation for Assigned Names and Numbers) in 1998, a non-governmental body which took control of the DNS (Domain Name System). The DNS's role is to assign domain names for websites and to link those to IP addresses. This seems a rather perfunctory, technical role but actually carries with it a lot of political power. This is because the Internet would not function without the DNS's capacity to police the naming of domains and their subsequent translation into IP addresses; setting up a replacement or rival system would be far too complicated and costly. The DNS thus constitutes what is referred to as the 'root' of the Internet and, by extension, whoever is responsible for the DNS thus effectively runs the Internet (Chadwick 2006: 234–235, 238). Thus, Internet users all over the world are subject to a regulatory framework that is drawn up in one country alone, the USA, whose government, despite the formal status of ICANN, effectively still controls the DNS and the root servers (Siapera 2012: 242; Unwin 2013: 541).

The increasing commercialization of the Internet has meant that this control brings with it enormous financial clout in addition to the political power mentioned above. Chadwick (2006: 244) highlights numerous examples of the legal and financial tussles that have taken place for the acquisition of highly prized domain names. In order to better adjudicate in these disputes, ICANN works with global trade and copyright bodies like the WTO and WIPO (World Information Property Organization). Its collaboration with the latter body has not been without controversy, as its decisions in intellectual property disputes have tended to favour large commercial copyright holders (Chadwick 2006: 218, 254). This is part of a more stringent global regime of intellectual property protection enforced by the WTO and WIPO. While this can protect those who work in the creative industries from copyright violations, it can also lead to a form of 'information feudalism' which enables large multinational corporations to patent 'indigenous' or 'folk knowledge' in less developed areas (Miller 2011: 69). An example of this is the patenting of the neem seed, a natural pesticide used in India, by an American multinational, with the consequence that indigenous farmers must now pay for a product which for years they have been using for free (Miller 2011: 69–70). While it would be wrong to argue that the WTO and WIPO are simply tools of the US government, it is fair to say that indigenous people in the developing world are not only not adequately represented on these bodies, but that they also do not have the financial means to fight expensive court cases against multinational corporations. ICANN's relationship with these bodies suggests that in this and many other aspects of Internet regulation it is perceived as not being on the side of the developing world.

Like the NWICO debates in the 1970s, as the Internet spread further around the globe, non-governmental bodies and developing nations became increasingly vocal about both the digital divide and the USA's effective sole control of cyberspace. Unlike the NWICO debates, these demands led to the organizing of two summits: one at Geneva in 2003 and one at Tunis in 2005 under the aegis of WSIS (World Summit on the Information Society). Organized by the ITU, WSIS was an attempt at a multi-stakeholder approach to Internet governance, featuring the private sector and representatives of global civil society alongside government officials. Not surprisingly, in a global gathering of this size and with such a wide array of actors, forging a consensus was difficult. Perhaps the major organizational problem centred on the respective representation of the private sector and civil society, with some commentators arguing that the former was given more weight than the latter (Sarikakis 2006: 167–168; Unwin 2013: 540). Similar to the gulf between proponents and opponents of NWICO in the 1970s this reflected an ideological difference between those who want to use ICTs for promoting a 'social welfare approach' and those who adopt an 'e-commerce approach' (Chadwick 2006: 225). Many in global civil society who championed the former approach set up the World Forum on Communication Rights (2003), an event which ran as an alternative to the 2003 WSIS summit (Chadwick 2006: 221). But even within the WSIS summit itself, there was public discord in the form of a 'Civil Society Declaration' by participants who felt that their concerns were not adequately addressed:

> At this step of the process, the first phase of the Summit, Geneva, December 2003, our voices and the general interest we collectively expressed are not adequately reflected in the Summit documents. We propose this document as part of the official outcomes of the Summit.
>
> (WSIS Civil Society Plenary 2003)

The Declaration called for a fundamental review of intellectual property regimes to include WIPO's role, and *inter alia* argued for the protection of indigenous forms of knowledge from patenting and the promotion in the developing world of free software for essential ICT functions (WSIS Civil Society Plenary 2003: 16–17). It also tackled the thorny issue of the global governance of the whole Internet, suggesting that:

> only a truly open, multistakeholder, and flexible approach can ensure the Internet's continued growth and transition into a multilingual medium. In parallel, when the conditions for system stability and sound management can be guaranteed, authority over inherently global resources like the root servers should be transferred to a global, multistakeholder entity.
>
> (WSIS Civil Society Plenary 2003: 22)

The official outcome documents from the summits in 2003 and 2005 barely mentioned intellectual property and did not suggest that its existing global regulatory regime was in any way flawed. It did, though, recognize that the existing arrangements for the governance of the whole Internet needed improvement. To this end, it proposed the setting up of the Internet Governance Forum (IGF). But any pretensions it had of replacing ICANN had already been ruled out by the US government (Siapera 2012: 242) and this was reflected in the admission that the IGF's role would be nothing more than advisory:

> *The IGF* would have no oversight function and *would not replace existing arrangements, mechanisms, institutions or organizations*, but would involve them and take advantage of their expertise. It would be constituted as a neutral, nonduplicative and non-binding process. *It would have no involvement in day-to-day or technical operations of the Internet* [my emphasis].
>
> (WSIS 2005: 84)

A background note on the IGF's website confirms its somewhat passive role: 'The IGF has no power of redistribution, and yet it has the power of recognition – the power to identify key issues' (IGF 2011). ICANN itself has introduced modest changes to its composition, including the global election of some representatives, but nothing substantial to justify a newspaper headline in 2009 to the effect that the 'US relinquishes control of the Internet' (Johnson 2009; Rossiter 2006: 89–90); indeed, the same newspaper, *The Guardian*, reported ICANN's chief executive Fadi Chehadé as saying in the wake of Edward Snowden's revelations about NSA surveillance that:

> it has always been envisaged, including written into the founding agreements, that the special relationship between Icann and US government will become more global *in the future*, and less focused on one government. So there's nothing new here [my emphasis].
>
> (Gibbs 2013)

While the Internet, for all its complexity, has run smoothly under ICANN, there is always the potential for the misuse of power if control is vested in only one nation-state.

> It doesn't take much to imagine the devastating effect of removing a CCTLD [country code Top-Level Domain name system management] in times of military and information warfare: a country's entire digital communications system is rendered useless in such an event, and social and economic impacts would come into rapid effect. A more likely scenario would see the US government intervening in the

allocation of CCTLDS in instances of political or economic dispute. In this regard, the control of top-level domain names operate as a potential form of economic sanction or a real technique of unilinear leverage in business and political negotiations.

(Rossiter 2006: 82)

But it is not only the United States that threatens the smooth running of the Internet. While the reaffirmation of its support for Article 19 of the UN's Universal Declaration of Human Rights makes it questionable whether, as one commentator intimates, commitments to communication-based human rights like freedom of expression were watered down after the Tunis summit (Flew 2007: 200–201; WSIS 2005: 57), the statement that 'Policy authority for Internet-related public policy issues is the sovereign right of States' could be interpreted as providing justification for state censorship (WSIS 2005: 76). Therefore, while it is not ideal for one country to effectively run the whole Internet, at least the USA's libertarian ideology and constitution makes it much more robust in defending freedom of expression than a multinational regulation regime in which countries like China and Russia would be highly influential would be likely to be.

Some analytical problems with traditional approaches to the digital divide

While there are very few who would argue that the digital divide between developed and developing nations should not be narrowed, many commentators question the high priority that governments, development agencies and activists place on it; one commentator states that, at the turn of the millennium, bridging the digital divide was the number one developmental priority for the World Bank (Wade 2004: 186). The first observation to make is that the digital divide appears largely to mirror existing disparities in levels of economic and social development (Chadwick 2006: 63; Miller 2011: 96; Wade 2004: 186). This correlates with the slowing down of growth in ICT capacity in the world's poorest nations highlighted earlier in the chapter; rapid growth levels in the late 1990s and the early years of the next decade could be explained by the acceleration that is always marked when one is starting from a very low economic base. This position, which is probably the most common in relation to the digital divide, supports the argument that the focus of all those interested in development should be on tackling these lags in social and economic development, rather than fixating on ICTs: at best, focusing on the latter is a distraction; at worse, it prevents activists from engaging in a necessary political critique of global inequality (Fuchs 2008: 223–224; Sarikakis 2006: 174–175).

In contradistinction to the instrumentalist approach above is one which is more technologically determinist in tenor. The discussion of Castells's theory of the network society in Chapter 4 suggested that he employs a kind of soft determinism. And it is undoubtedly the case that his network society thesis, with all that it implies socially and economically for those who are marginal to or excluded from it, has had a significant influence on debates about the digital divide (Huyer 2006: 213–214). This soft determinism appears to imply that many social and economic problems can be overcome with the use of ICTs and the rhetoric of some of the champions of the role of ICT in development veers dangerously near to technological determinism. That said, a blanket denial of any link between a society's level of technological development and its social and economic status does run the risk of blaming the citizens themselves for their societies' relative lack of development (Burton and Van Der Hoven 2007). This needs to be borne in mind in the discussion on the examples of *use* of ICT in the developing world which closes this chapter.

Another aspect, perhaps, of the influence, not only of Castells, but also other theorists and policy-makers who champion the information society is in the tendency of some commentators to make what is a complex and multi-faceted problem almost solely reducible to economics (Siapera 2012: 69). This explains the prominence of the private sector at the WSIS summits but, as this author has argued throughout this book, more broadly reflects the position of neo-liberalism as the global political ideology without peer. The WSIS outcome documents published in December 2005 are replete with references about the need to develop e-business, in particular the building of SMMEs (small, medium and micro enterprises), and the championing of public/private partnerships as the most effective delivery for ICT development; as noted by others above, this author could not find one instance in the lengthy document where the 'private sector' was not placed immediately after 'governments' and before 'civil society' (WSIS 2005). The problem with this type of approach is that it rules out any alternative form of social and economic development.[30] A failure to acknowledge that it is the existing global economic system which is part of the problem means perversely that for some nations narrowing the digital divide will make them more, rather than less, dependent on governments and corporations in the developed world. How so? The dominance of corporations from the developed world in the computer hardware and software market means that they will benefit greatly from the creation of new markets for their products in the developing world. The developing world will also be more dependent on 'western expertise' to help them fully benefit from these technologies (Wade 2004: 185–186). In this sense, it is clear that the private sector representatives at WSIS have a vested interest in advocating the narrowing of the digital divide as the most effective form of development. It is worth relaying at length here the reported views of Intel Corporation's John Davies at the 2012 WSIS forum (though no summit has been held since 2005, WSIS continues its work in fora):

> As a result of successful collaborations, Intel recently launched the programme called Reaching the Third Billion. Since only two billion people owned PCs and had access to the Internet, the idea was to bring the programme to the third billion. The concept was to replicate the miracle that had happened with prepaid cell phones in the past, and apply it to the PCs and other devices in broadband. Intel is working with several countries, along with the Broadband Commission's support, typically under the leadership of Dr. Toure, ITU Secretary-General.
>
> Mr. John Davies shared that in 2011, at the UN headquarters in New York, Intel announced that they would undertake a training programme to train teaches [*sic*] to use PCs, he announced that through this programme, they had trained 11 million teachers to use PCs in classrooms. The purpose was to consider whether some of those techniques would also work for health care workers. As a result, it was agreed that by 2015 Intel would potentially train a million frontline health workers in emerging economies on ICTs.
>
> (WSIS Forum 2012: 13)

While Intel's training programmes may indeed benefit people in key development sectors in the world's poorest economies, it is doubtful whether the company is providing them for solely charitable reasons: in the horrible phraseology of the corporate world (which, unfortunately, infests the modern university too!), the company 'congratulated the WSIS Forum as a platform for building *win-win partnerships*' [my emphasis] (WSIS 2012: 13). Though it proved less than successful, Nicholas Negroponte's programme to deliver $100 laptops to children in the developing world would have made recipients' dependent on companies like Microsoft, who supply the necessary operating systems and software. What made this programme even more problematic was that the laptops were inferior to those sold throughout the developed world and so would have lumbered millions of people with systems which would work frustratingly slowly (Fuchs 2008: 221). A move towards this type of ICT dependency also ties users in the developing world into global intellectual property regimes from which they gain little benefit.

The final criticism of conventional scholarly approaches to the digital divide stems from the empirical foundation on which they are based. As Chadwick (2006: 51) notes, discussions of the digital divide tend to focus on physical access (or lack thereof) to ICTs, which is measured by the kind of statistics which have been included in this chapter. This type of approach is useful in providing evidence that the digital divide actually exists, but is inadequate in determining how factors like technical skills and social status affect engagement with ICTs; in many parts of the developing world, for instance, women's access to the Internet is restricted by discriminatory social conventions (Unwin 2013: 533), a phenomenon that might not be recognized if one employs crude statistics alone. The use of statistics also encourages the identification of patterns and the proving of causation through correlation. But, as Chadwick (2006: 63–67) and Selwyn (2004: 357–358) both point out, the way in which ICTs are used often confounds these neat patterns and correlations. To gain a better appreciation of people's engagement with ICTs, it is necessary to think about how they are actually *used* and this chapter will end with some relevant examples of the use of ICTs from different parts of the developing world.

The importance of use – how ICTs are used in various parts of the developing world

An example of a debate about engagement with digital media in the developing world which often fails to engage adequately with theories of use is the one about the so-called Arab Spring. The political unrest in many Arab countries at the beginning of the second decade of this century could be said to have begun in Iran in 2009.[31] Like the civil unrest that followed in other countries in the region, digital media appeared to play a predominant role in recording and disseminating abuses carried out by the state's security forces, as well as coordinating the protests against the 2009 election results (Rahimi 2011). *The Washington Times* used a term which subsequently became a popular means of encapsulating the role of micro-blogging in the protests: 'The Twitter Revolution' (Mason 2012: 35). YouTube was also an influential player in the drama that unfolded, particularly in its disseminating of the horrific footage of the killing of protestor Neda Agha-Soltan by the Basij militia on 20 June 2009 (Mason 2012: 35–36). Though this particular revolution failed, the use of social media in protests in the Arab world over the next few years met with some success.

In Tunisia, images of the self-immolation of fruit-seller Mohamed Bouazizi in December 2010 in despair at his harassment by local officials were quickly put on Facebook, and the outrage led to mass protests which eventually brought down President Ben Ali (Khondker 2011: 676–677; Mason 2012: 32). Around the same time Egypt faced similar revolutionary upheaval, where the beating to death of a political blogger in June 2010 led to the creation of a Facebook group named after him: 'We are all Khaled Said'. Set up by Google employee Wael Ghonim, this provided both a focus for and a means by which to organize the protests that eventually brought down Hosni Mubarak (Faris 2013: 3; Hill 2013: 65). Varying degrees of unrest and protest spread across the region, including a bloody revolution with the aid of foreign intervention in Libya and (at the time of writing) continuing civil war in Syria.

The significance of the role of various forms of digital media in these uprisings has been keenly contested by scholars, with the almost-technological-determinism of those who proclaim 'Twitter' and 'Facebook' revolutions being countered by those for whom the deployment of these digital media technologies had little more than a marginal effect, or indeed were arguably more effectively used by authoritarian regimes than by progressive activists (Mason 2012; Morozov 2013; Shirky 2011). It is common for analysts to approach the same issue in these diametrically opposed ways, and for others to take a more subtle position which, like the theory of media use expounded throughout Part III of this book, emphasizes the complexity and variability of the interaction between these media and human beings (Axford 2011; Christensen 2011; Comunello and Anzera 2012; Faris 2013; Khondker 2011). But all too often the debate is characterized by crude models of causation based on relative rates of access to digital media and the empirical outcomes of the uprisings. Thus, there is a tendency even for relatively sophisticated accounts of these uprisings to write that 'part of the reason' similar protests in 2008 failed to trigger mass protests 'was that at time there were only 28,000 Facebook users in Tunisia' (Khondker 2011: 677). Celebratory accounts of the role of digital media in these protests stress the rapid take-up of social media immediately prior to them (Mason 2012: 135), while those of a more sceptical bent emphasize the tiny percentage of Twitter users located in Egypt and Tunisia (Curran 2012: 51–52). Using statistics to support theory-based research is fine; it would be odd if an author whose work in this book is replete with statistics would say otherwise! Nonetheless, debates about digital media and causality in the Arab Spring would benefit from more primary research in the countries themselves in order to identify the precise ways and extent to which these media were actually *used*, something which David Faris's book-length study of the Egyptian revolution advocates:

> mobilizations triggered by Social Media Networks appear to have no determinative relationship to the success or failure of activism in a given authoritarian context. The divergent outcomes of SMN-driven activism in Kenya, Ukraine, Moldova and Iran demonstrate that scholars and observers cannot simply assume the democratizing impact of SMNs. Indeed, in Kenya, the technologies were put to much more nefarious uses, whereas in Iran SMNs were turned on their users quite effectively. The success of Ukrainian, Moldovan and ultimately Egyptian SMN activists should be studied further for better understanding of the dynamics which allowed these groups to succeed.
>
> (Faris 2013: 200–201)

In Africa, there have been many studies of digital media use, many of which challenge the bleak picture that the statistics alone present. The earlier point about it being extremely common in Africa for people to share mobile phones, as well as to use intermediaries to access them, illustrate how basing analyses solely on statistics on the one device/one user model over-estimates the cost per person, while under-estimating individual use. Without wishing to downplay the problems that still exist in Africa, knowledge of the way in which digital media is actually used suggest sophisticated methods of increasing access which the statistics cannot capture. Themba Mbhele, the General Secretary of the Anti-Privatisation Forum, a South African NGO, describes how his organization gets its message across in spite of considerable technological obstacles:

> Most of our comrades do not have access to computers or cell phones. Specific comrades in the community are contacted through cellular phones by head office and from there the message is sent to all in the community. We are putting up more offices in the townships

which bring the organisation closer to the people. Everybody is thus well-informed regardless of computer access.

(Rolls and October 2004; cited in Wasserman 2007: 141)

Indeed, good, evidence-based policy-making must take actual use into account when formulating digital media programmes in the developing world. This clearly did not happen when multi-purpose telecentres were established in various parts of rural South Africa and Mexico to improve access to the Internet. While, in the case of Mexico, the World Bank assumed that there were local impediments to their effective working, it is more likely the case that the programme failed in both countries because it did not take users' actual needs into consideration (Burton and Van Der Hoven 2007; Wade 2004: 187–188).

Conversely, there are occasions where a successful bridging of the digital divide can cause more problems than benefits. It could be argued that this is the case with some of the mobile banking transactions that were mentioned earlier in this chapter. These do have undoubted benefits for people in the developing world, especially in allowing farmers to check commodity prices and in enabling migrant workers to remit money home; Miller cites an *Economist* article in 2009 which reported that seven million Kenyans out of total population of 38 million used the mobile money transfer scheme M-PESA, trading around $2 million every day (*The Economist* 2009; cited in Miller 2011: 65). But this also binds people in developing nations more closely to the international capitalist system (Cere 2006; Poster 2006: 83–84; Unwin 2013: 539). While Castells would argue that it is counter-productive to attempt to opt out of the global network society, when that network facilitates flows of capital out of Africa which massively dwarf the financial inflows of ordinary Africans using schemes like M-PESA then there are too clearly dangers associated with imbrication in the network (Shaxson 2011).

As the world's most populous country and second largest economy, China is in the unique position of being a developing nation with immense power. Thus it has the capacity to forge an alternative path in the development and use of digital media technologies. What shapes a lot of digital media use in China is the looming presence of the state. It is well known that the Chinese state is highly censorious and this practice extends inevitably to digital media. In many respects this is what one would expect from an authoritarian, one-party regime. In addition to these naked political reasons for restricting content, it could be argued that there is an element of economic protectionism in play. The overseas social media platforms and search engines that are blocked in China all have Chinese equivalents: micro-bloggers prevented from using Twitter can use Weibo; restrictions of Google lead many to use Baidu instead; if you cannot watch YouTube, you can watch Youku; and who needs Facebook when you have Renren? This has led some to argue that the motivation for blocking or restricting these overseas platforms is partly economic (Anti 2012). The same commentator has also highlighted the need for the Chinese government to own the databases which store social media activity by Chinese citizens (Anti 2012). Edward Snowden's allegations about the NSA trawling the databases of American-based Internet companies suggest that the Chinese government's (or any other government's for that matter) fears are not entirely without foundation. Also, economic protectionism for burgeoning media companies in the developing world was recommended in the 1980 MacBride Report. What the case of China illustrates, though, is that however imperfect the Anglophone libertarian model is, constructing a viable alternative is not an easy task.

The power of the Chinese political system conditions even resistance to it. An example of this is the peculiarly named phenomenon 'human flesh searching', or *ren rou sou suo* in Chinese. This involves the hunting down of individuals who Internet users have deemed to have transgressed in some way. One of the earliest high-profile examples of human flesh searching involved the identification of a woman who in 2006 stabbed a kitten to death with her stiletto heel (Cheong and Gong 2010: 472). While it is not uncommon for people to be hunted down for personal transgressions, which include extra-marital affairs, human flesh searching is arguably used mainly to expose the wrongdoing of government officials (Cheong and Gong 2010). While this can be read as a relaxation of the strict censorship regime, a more subtle interpretation of it is that the central government tolerates, if not downright encourages, the scrutiny of local officials as a means of cracking down on corruption, which some see as one of the biggest threats to the central leadership's legitimacy (Cheong and Gong 2010). This is similar to the campaigns by Mao Zedong in the 1960s and 1970s to turn the masses against officials and hence should caution one against thinking that human flesh searching is simply a manifestation of greater openness on the Chinese Internet (Downey 2010). The observation that campaigns against leading national political figures have not fared so well reinforces this argument.[32] Nonetheless, there are opportunities for much bolder critics to get their message across. The Chinese language has the capacity to produce many more homonyms than the Roman alphabet and critics of the central government often take advantage of this. An example of this is the subversion of the word for 'harmony', a metaphor for the type of stable society which Chinese leaders believe characterizes modern China and which they argue political dissidence threatens. The pinyin for harmony is 'he xie', a word which can also mean 'river crab' (Wang 2012). In this sense, criticisms of the concept of harmony can use the latter term to evade censorship. After a while, this type of in-joke might be identified by the censors, but by then new phrases might have been invented. Another important aspect of the Chinese language in relation to digital media is the fact that most Chinese words are expressed in two characters. This means that the Chinese equivalent of Twitter, Weibo, enables user to write much more using 140 characters than the platform which is dependent on the Roman alphabet. Weibo also has the capacity to send images and videos, functions which are not available to users of Twitter. This seems to have provided an outlet for critical views, though whether or not this merely provides a safety-valve to enable people to let off steam without undermining the political system is a moot point (Hewitt 2012).

As with the enthusiasm among many western commentators to ascribe revolutionary properties to Twitter and Facebook in the Arab world, it would be a mistake also to view digital media use in China through a similar prism. While a World Internet Project survey reported that a much higher proportion of Chinese respondents (46 per cent) than other nationalities believed that the Internet 'will give users more of a say in government actions' (WIP 2008; cited in Cardoso, Liang and Lapa 2013: 228), it appears that the Chinese use digital media mainly for the purpose of entertainment. The proportion of its users who play games online, watch videos and listen to music is much higher than in India, Brazil, Russia, Indonesia, Japan and the USA (Michael and Zhou 2010: 9); in this regard, concerns about high levels of Internet addiction (often linked to gaming) can be deemed to be more than simply a manifestation of censorious behaviour by the Chinese government (Unwin 2013: 548). That the country's most popular Weibo user is actress Yao Chen is further evidence that digital media use in China is oriented more towards entertainment than politics (Osnos 2013).

Cultural imperialism in reverse? The developing world's export of digital media practices to the developed world

In the previous chapter, Juris (2008) reported how the Zapatistas not only gained worldwide support for their particular cause, but also introduced a new methodology for Internet-based political campaigns which was exported from the Chiapas region of Mexico to many political struggles in the developed world. This is an example of the way in which cultural imperialism can work in reverse, where globalization provides opportunities for people and movements in the developing world to export their expertise to the developed world. Due to its greater functionality than Twitter, Chinese micro-blogging platform Weibo is attracting many more people from the developed world (DeWoskin 2012), and I know from my own experience with my friends and family that an increasing number of non-Chinese people are signing up to social networking platform Weixin or WeChat. The most successful of these exports from the developing to the developed world, though, arguably, is *Ushahidi*, and this chapter will end with an account of its rise.

Meaning 'witness' or 'testimony', *Ushahidi* was developed in Kenya to identify locations where violence was taking place in the aftermath of the December 2007 elections. The real-time maps that were produced from crowd sourcing information were much more accurate in their reporting than official media outlets in Kenya, where passivity or partisanship prevented them from providing citizens with useful information about the outbreaks of ethnic violence (Shirky 2010: 15). One of the most significant features of *Ushahidi* was that it received information in the form of text messages, rather than from Internet sources (Unwin 2013: 545). If we recall that east Africa's access to the Internet was patchy until submarine fibre-optic cables were laid along its coast in 2009, then this is a great example of an innovation which is not restricted by limited access to broadband technology. Such was the success of this platform that the technology has been exported to other parts of the developing world, being used for the same reason that it was used in its original application in the Democratic Republic of Congo, to combat voter fraud in India and Mexico, and to help find injured people in earthquakes in Haiti and Chile (Shirky 2010: 16–17). But it was its use in mapping crisis points during the 2011 Japanese earthquake and snow clean-ups in New York City that demonstrated how developing world technologies can work successfully in the developed world too (*MIT Technology Review* 2013; Talbot 2011). It is heartening to see that Kenya, a country that ITU statistics ranks only 116 out of 157 countries in terms of ICT development, can create not only *Ushahidi*, but also the earlier-mentioned international payments system M-PESA, as well as *Huduma*, an electronic system to improve governmental transparency (*BBC News* 2013b; ITU 2013: 24). Kenya is thus an example *par excellence* of how the innovative use of available technologies can go some way to overcoming the digital divide.

Epilogue

My reluctance to indulge in futurology means that any reader expecting a prediction of future trends in this final section will be disappointed. I also do not intend to provide a lengthy recapitulation of themes that have been extensively covered in the previous chapters. Nonetheless, I think it would be useful to finish by highlighting the most distinctive ways in which digital media exert influence in the political, economic and social spheres. Given the emphasis throughout on case studies as a means by which the rich variety of digital media use can be explored, the book will end with a discussion of the use of digital media in the dramatic and tragic days which followed the bombing of the 2013 Boston marathon.

In April 2013, surgery to remove a blood clot prevented me from flying to Cambridge to present at the Media In Transition conference at MIT University. My disappointment at missing this event was soon eclipsed by more important concerns in the form of news about the fatal shooting of a police officer at the campus, just days after three people were killed in an explosion at the Boston marathon. I found out about the shooting on the afternoon of Friday 19 April, China time. As I was still recuperating from my operation, I had time to follow the search for the suspects on social media. Though I was hampered by the restrictions imposed on English language social media platforms in China, I was able to find enough websites to enable me to be kept up-to-speed with the latest developments. In what was effectively an investigation by crowd sourcing, digital media users sifted through photographs taken on the day of the Boston marathon and quickly identified a number of suspects. Cross-referencing this visual evidence with information that had been reportedly picked up on the Boston Police Department scanner, the online search was quickly narrowed to two suspects, Sunil Tripathi and Mike Mulugeta (Madrigal 2013). As quickly became apparent, this information was completely wrong. In the case of Tripathi, a missing Brown University student whose body was discovered less than a week later, this unwanted publicity caused further agony to a grieving family (Buncombe 2013; Williams 2013); in relation to the other 'suspect', it is not even clear if a person matching 'his' description actually exists. While two more credible suspects were quickly identified – one, Dzhokhar Tsarnaev, who was on trial for the murders as this book neared completion, and the other, his brother Tamerlan, who was killed during the police manhunt – this did not lessen the damage caused by the misidentification of Tripathi and Mulugeta as suspects. This can be gauged by a simple Google search: when I searched for Mike Mulugeta's name, by the time I had typed in 'Mike Mulu...' I was prompted to complete my entry with 'mike mulugeta sunil tripathi'.[33] Worse is the Google Image search for the same names, where Tripathi's image (there do not seem to be any images of 'Mike Mulugeta', reinforcing the suspicion that a person fitting his name and description does not actually exist – see Leys 2013) is interspersed with photographs of the Boston bombing scene including photographs of suspects who look like the Tsarnaev brothers.[34] It is worth noting that this 'information' exists in the public domain long after a public *mea culpa* by one social media site, whose users were among the most active in misidentifying the suspects (*Reddit* 2013), and the closing of the thread, 'findbostonbombers', in which most of the speculation on its website took place.[35]

These inaccuracies did not appear to stem the flow of information on social media platforms during the manhunt for Dzhokhar Tsarnaev that unfolded in the hours following the fatal shooting of MIT police officer Sean Collier. This included the posting online of photographs taken by residents of Watertown, the district of Boston where the police were searching for Tsarnaev, some of which showed police and military personnel (DeNinno 2013). While CNN put its 'live' coverage on tape-delay at certain points of the search lest it broadcast violent scenes to its audience (probably following the advice of the police), there was no such self-denying ordinance by users of social media. Images of the precise position of police officers and soldiers were displayed online despite the Boston Police Department's plea on Twitter that such information could compromise the safety of the searchers (*The Washington Times* 2013). Similarly, it seems that US police do not have a mandate to prevent websites like *Broadcastify* providing live audio streams from police scanners;

Broadcastify did, however, suspend its operations during the manhunt for Dzhokhar Tsarnaev (Kelly 2013).

We can link aspects of the coverage of the Boston bombings and the hunt for those responsible with the main themes of this book. New knowledge is being rapidly created and disseminated globally by digital media. As Part I argued, this can be useful in challenging elitist views of the world, and has been used sometimes to great effect by the alter-globalization movement. However, as the case of the Boston marathon bombing and its aftermath show, these new techniques, especially those which are based on crowd sourcing all too often validate Mark Twain's oft-quoted observation that the 'a lie can travel half way around the world while the truth is putting on its shoes' (cited in Kaid 2000: 147). From a libertarian perspective, some have argued that this is best resolved by news producers and users themselves adopting a 'slow news' approach, which would involve a more deliberative and thoughtful way of sifting accurate from inaccurate information (Gillmor 2010: 25–29). However, this argument is problematic in two respects. The first is that it ignores the basic political economy of global media. As argued in Part II, the global economy is increasingly predicated on speed of transaction and of communication, and hence it is difficult to imagine successful media corporations prospering by allowing their rivals to beat them to the scoop, or users of social media *en masse* to refrain from passing on 'information'. Secondly, while these inaccuracies are often refuted – in the case of Tripathi and 'Mulugeta' within one day – inaccurate information stubbornly endures online long afterward and, as the examples above have shown, even persist near the top of Internet searches. This is not the fault of Google or any other search engine, but the problem is exacerbated by the existence of personal information in databases long after it has ceased to be useful. Mayer-Schönberger (2009) has argued that each piece of personal information online should have a legally enforceable fixed expiry date, and I would argue that, with regards to the retention of information, it would be better to put the onus on deletion rather than retention and apply it right across the board to cover commercial as well as governmental databases. My final point is about users, the subject of Part III of the book. While surveillance-from-below can sometimes be useful in holding governments to account, more often than not, as my examples have shown, it is ordinary citizens who suffer the indignity of this intrusion. Even where agents of the state, like police officers, are caught on camera behaving inappropriately, judicial systems even in democratic countries seem almost as reluctant to prosecute them as they were in the days when this behaviour was not recorded. It is not my intention to end this book on a negative note; as I have shown throughout Part III, citizens are using digital media for progressive reasons in ever more inventive ways. But while celebrating these uses we should also not ignore the political and economic systems which shape the engagement between digital media and ourselves.

Notes

1   From the Public to the Private: The Digitization of Scholarship

1. This interpretation is contested by Battles (2003: 31), who argues that later generations retrospectively conceived this origin myth in order to justify their own proselytizing of universality. This theory is supported by the knowledge that contemporary accounts were extremely vague about the layout of the library (Bragg *et al.* 2009; Polastron 2007: 12). But, given that most Alexandrian scholars assume that this was a credible attempt to create a universal library, the section will proceed from this premise; in any case, even if this motivation has been ascribed retroactively, its *function* in encouraging subsequent attempts is more important than absolute proof of its veracity.
2. The plausibility of this explanation is challenged by Cavallo and Chartier's assertion that it was the thirteenth-century mendicant Orders that introduced functionality to the catalogue: 'No longer a simple inventory, the catalogue became an aid to consultation that noted where books could be found in a given library or even in a given geographical area' (1991: 19). (This implies that Callimachus's catalogue did not have the same functional character.) Whatever the truth about these competing claims, it is reasonable to view Callimachus's work practically as an antecedent of our contemporary catalogues.
3. Crowd sourcing is a practical embodiment of the credo 'the wisdom of crowds'. It works by inviting users to identify textual errors and suggest corrections, a methodology employed by the National Library of Australia for its huge newspaper digitization programme (NLA 2013).

2   From the Private to the Public: Online Identity

4. Web 2.0 refers to digital media technologies which are more interactive than previous, supposedly more static platforms, and in that sense is associated with the growth of social media since the middle of the first decade of the twenty-first century.
5. See   http://www.pbs.org/wgbh/pages/frontline/digitalnation/relationships/identity/the-secret-online-life-of-autumn-edows.html.

3   Digital Media and Politics in the Liberal Democratic State

6. Thanks to Phil Ramsey for drawing my attention to this quotation, as well as for his general comments on this chapter.
7. For a discussion of how a digital commons might be created, see Ramsey (2013).

5 The Digital Economy and the Global Financial Crisis

8. As this book was being completed, an International Labour Organization (ILO) report highlighted how recent worldwide economic growth was not being matched by a commensurate growth in employment (Fowler 2014).
9. Anita Elberse's research for her first book *Blockbusters*, published at the end of 2013, supports this thesis (Stevenson 2014).

6 Reading/Using Digital Media

10. *Victory Garden* is available here: http://www.eastgate.com/VG/VGStart.html.

7 The New Social Movements

11. The term was coined by Richard Barbrook and Andy Cameron in the 1990s (Barbrook and Cameron 1995; Turner 2006: 208).
12. It is not clear, even after reading the vast literature on the subject, what is the difference between social movements and *new* social movements. For the purpose of clarity of argument in this chapter, social movements will refer to these groups generally, whereas the term 'new social movements' will be used specifically to conceptualize those same movements' utilization of digital media for their activism. While the distinction is primarily a conceptual one, inevitably there is a temporal differentiation too: put simply, the activities of these social movements from the advent of the World Wide Web in the early 1990s onwards will be referred to with the 'new' moniker.
13. I find the term 'anti-globalization' too negative, and an inaccurate way of characterizing this type of activism, so, following Jordan (2007: 75), I will use the term 'alter-globalization'.
14. Indymedia has websites devoted to many cities around the world, but a more general, global portal can be accessed here: http://www.indymedia.org/or/index.shtml.
15. The *G20 Meltdown* website no longer exists but its URL was: http://www.g-20meltdown.org/.
16. As with the example of *The Observer's* criticism of the demonstrators leading up to the march, even left-leaning newspapers like the *Daily Mirror* were not averse to uncritically adopting the police's story of an out-of-control mob (Gregory 2009): http://www.mirror.co.uk/news/uk-news/man-dies-in-g20-protests-385819.
17. See this report in Chicago's *Daily Herald*: http://prev.dailyherald.com/story/?id=394348.

8 Surveillance: The Role of Databases in Contemporary Society

18. Though the official date of publication is 2007, save for a few minor changes, this was a version of a chapter which appeared in a monograph in 2001; Manovich reveals that, though published in 2001, the chapter was actually written in 1998 (Manovich 2007: 58).
19. Available at: http://www.zoominfo.com/p/Andy-White/734713281.

20. Available at: https://ningbo.academia.edu/AndrewWhite.
21. I thank David Fleming for recommending reading sources for the section on Deleuze's control society.
22. For all we know, it could still be in existence.
23. See a succession of reports from August to November 2008 on the BBC's website: *BBC News* 2008a: http://news.bbc.co.uk/2/hi/uk _news/7575766.stm; *BBC News* 2008b: http://news.bbc.co.uk/2/hi/uk_news/ england/7619177.stm; *BBC News* 2008c: http://news.bbc.co.uk/2/hi/uk _news/politics/7667507.stm; *BBC News* 2008d: http://news.bbc.co.uk/2/hi/ uk_news/7704611.stm.
24. The successful prosecution in 2013 of a British soldier for murder in Afghanistan was largely the result of evidence from a six-minute video of the killing filmed by one of the other soldiers (Morris 2013).

9   Digital Media Use in the Developing World

25. Though I currently work in China, it is for a UK university.
26. See http://www.google.co.uk/search?q=%22african+people%22&new window=1&source=lnms&tbm=isch&sa=X&ei=b2jDUqbnJoqriAeT2YG4DA& ved= 0CAcQ_AUoAQ&biw=1366&bih=664.
27. This chapter will sometimes use the term 'ICT' rather than 'digital media', as this is the preferred term when activists, policy-makers and scholars refer to these technologies in the context of the developing world.
28. The Y2K or, as it was popularly known, the millennium bug, stemmed from the fact that internal computer clocks recorded only the last two digits of the year. Therefore, the year 2000 would be recorded as 00, and it was feared that systems might close down as the computers might read this as the year 1900. It is not clear whether the problem was ever that serious but it no doubt created a lot of work for IT professionals!
29. Miller (2011: 107) goes on to point out that even a country as advanced as France did not have what would be considered a 'modern' fixed-line infrastructure until the 1970s.
30. An example of a radical alternative approach to bridging the digital divide is Christian Fuchs's (2008: 223–224).
31. It should be noted that, though it shares the same main religion (Islam) as many states in the region, Iran is not an Arab country.
32. At the time of writing, the website of the *New York Times* is blocked in China as a result of it publishing a number of negative articles about Chinese leaders, including the former Premier Wen Jiabao.

Epilogue

33. See: http://www.google.co.uk/#newwindow=1&q=mike+mulugeta+sunil + tripathi.
34. See: http://www.google.co.uk/search?q=mike+mulugeta+sunil+tripathi& newwindow=1&tbm=isch&tbo=u&source=univ&sa=X&ei=CavsUpuYJ8Xyi AeFiICAAg&ved=0CCoQsAQ&biw=1366&bih=664.
35. See: http://www.reddit.com/r/findbostonbombers.

Bibliography

Aarseth, E. (1997) *Cybertext: Perspectives on ergodic literature.* Baltimore: Johns Hopkins University Press.

Aas, K.F. (2004) From narrative to database: Technological change and penal culture. *Punishment & Society*, 6(4): 379–393.

Academia.edu (2013) *Andrew White* [profile]. Available: https://ningbo.academia.edu/AndrewWhite.

Adler, J. (2012) Raging bulls: How Wall Street got addicted to light-speed trading. *Wired*, 3 August. Available: http://www.wired.com/business/2012/08/ff_wallstreet_trading/.

Alford, B. (1997) De-industrialisation. *Refresh*, 25: 5–8. Available: http://www.ehs.org.uk/dotAsset/921ccd8d-b69f-442c-8515-05063e5de4fb.pdf.

Anderson, C. (2007) *The long tail: How endless choice is creating unlimited demand.* London: Random House.

Anderson, C. (2009) *Free: The future of a radical price.* London: Random House Business Books.

Andrejevic, M. (2006) The discipline of watching: Detection, risk, and lateral surveillance. *Critical Studies in Media Communication*, 23(5): 391–407.

Anti, M. (2012) *Behind the great firewall of China. TED Talk.* Available: http://www.ted.com/talks/michael_anti_behind_the_great_firewall_of_china.html.

Appiah, K.A. (2007) *Cosmopolitanism. Ethics in a world of strangers.* London: Penguin Books.

Apple Press Info (2012) *Fair Labor Association begins inspections of Foxconn*, 13 February. Available: http://www.apple.com/pr/library/2012/02/13Fair-Labor-Association-Begins-Inspections-of-Foxconn.html.

Appleyard, B. (2007) Could this be the final chapter in the life of the book? *Sunday Times*, 21 January.

Aronowitz, S. (2001/1994) Technology and the future of work, in Trend, D. (ed.) *Reading digital culture.* Malden, MA, Oxford Carlton, Australia: Blackwell Publishing, pp. 133–143.

Associated Press/(Chicago) (2010) Authorities in Belarus disperse pillow fight . . . as national security threat. *Daily Herald*, 15 July. Available: http://prev.dailyherald.com/story/?id=394348.

Athique, A. (2013) *Digital media and society: An introduction.* Cambridge: Polity Press.

Auletta, K. (2010) *Googled: The end of the world as we know it.* London: Virgin Books.

Australian Museum (2013) *Nature culture discover.* Available: http://australianmuseum.net.au/Cane-Toad.

Axford, B. (2011) Talk about a revolution: Social media and the MENA uprisings. *Globalizations*, 8(5): 681–686.

Baker, N. (1992) *Double fold: Libraries and the assault on paper.* New York: Vintage Books.

Banderas News (2006) As tower fades at home, it still shines abroad, December 2006. Available: http://www.banderasnews.com/0612/ent-towerfades.htm.

Barbrook, R. (2007) *Imaginary futures: From thinking machines to the global village.* London and Ann Arbor, Michigan: Pluto Press.

Barbrook, R. and Cameron, A. (1995) The Californian ideology. *Mute*, 3(Autumn). Available: http://w7.ens-lyon.fr/amrieu/IMG/pdf/Californian_ideology_Mute _95-3.pdf.

Barron, R.M. (2008) *Master of the Internet: How Barack Obama harnessed new tools and old lessons to connect, communicate and campaign his way to the White House.* Swathmore, PA: Swathmore College. Available: http://web.cs.swarthmore.edu/ ~turnbull/cs91/f09/paper/barron08.pdf.

Battles, M. (2003) *Libraries: An unquiet history.* London: Vintage.

Baudrillard, J. (1995) *The Gulf War did not take place.* Bloomington: Indiana University Press.

Bauman, Z. (2000) *Liquid modernity.* Cambridge and Malden, MA: Polity.

Baym, N.K. (1998) The emergence of on-line community, in Jones, S.G. (ed.) *Cyber-society 2.0.* Thousand Oaks, California: Sage, pp. 35–68.

BBC News (2008a) Company loses data on criminals, 21 August 2008. Available: http://news.bbc.co.uk/2/hi/uk_news/7575766.stm.

BBC News (2008b) NHS memory stick found in street, 16 September 2008. Available: http://news.bbc.co.uk/2/hi/uk_news/england/7619177.stm.

BBC News (2008c) Up to 1.7m people's data missing, 13 October 2008. Available: http://news.bbc.co.uk/2/hi/uk_news/politics/7667507.stm.

BBC News (2008d) Probe into data left in car park, 2 November 2008. Available: http://news.bbc.co.uk/2/hi/uk_news/7704611.stm.

BBC News (2010) Identity cards scheme will be axed 'within 100 days', 27 May 2010. Available: http://news.bbc.co.uk/2/hi/8707355.stm.

BBC News (2011) Jody McIntyre: Met police sued over wheelchair complaint, 27 May. Available: http://www.bbc.co.uk/news/uk-england-london-13578369.

BBC News (2012) G20 death: PC Simon Harwood sacked for gross misconduct, 17 September. Available: http://www.bbc.co.uk/news/uk-19620627.

BBC News (2013a) Bomb threats on Twitter made against female journalists, 1 August. Available: http://www.bbc.co.uk/news/world-23523539.

BBC News (2013b) Kenya launches Huduma e-centre to cut bureaucracy, 7 November. Available: http://www.bbc.co.uk/news/world-africa-24855993.

BBC News (2014) Obama orders curbs on NSA data use, 17 January. Available: http://www.bbc.co.uk/news/world-us-canada-25785573.

Beck, U. (1992) *Risk society: Towards a new modernity* (translated by Mark Ritter). London: Sage.

Bell, D. (1973) *The coming of post-industrial society: A venture in social forecasting.* New York: Basic Books.

Bennett, C.J. and Parsons, C. (2013) Privacy and surveillance: The multidisciplinary literature on the capture, use, and disclosure of personal information in cyberspace, in Dutton, W.H. (ed.) *The Oxford handbook of Internet studies.* Oxford: Oxford University Press, pp. 486–508.

Berry, D.M. (2011) *The philosophy of software: Code and mediation in the digital age.* Basingstoke and New York: Palgrave Macmillan.

Biggs, J. (2013) Lavabit founder details government surveillance of secure email while documents disclose epic trolling of Feds. *Techcrunch.com*,

3 October. Available: http://techcrunch.com/2013/10/03/lavabit-founder -details-government-surveillance-of-secure-email-while-documents-disclose -epic-trolling-of-feds/.

Birkerts, S. (1994) *The Gutenberg elegies: The fate of reading in an electronic age.* New York: Ballantine Books.

Birnbaum, J.H. (2000) Politicking on the Internet. *Fortune,* 6(March): 86–88.

Boeder, P. (2005) Habermas' heritage: The future of the public sphere in the network society. *First Monday*, 10(9). Available: http://firstmonday.org/ojs/index .php/fm/article/view/1280/1200.

Bogard, W. (2009) Deleuze and machines: A politics of technology? in Poster, M. and Savat, D. (eds) *Deleuze and new technology.* Edinburgh: Edinburgh University Press, pp. 15–31.

Bomford, A. (1999) Echelon spy network revealed. *BBC News*, 3 November 1999. Available: http://news.bbc.co.uk/2/hi/503224.stm.

Bragg, M., Goodhill, S., Cuomo, S. and Nichols, M. (2009) *In our time.* London: BBC Radio 4, 12 March.

Branch, T. (2009) *The Clinton tapes.* London: Simon & Schuster.

Brook, S. and Gillan, A. (2009) MPs' expenses: How scoop came to light – and why journalists fear 'a knock on the door'. *The Guardian*, 18 May. Available: http://www.theguardian.com/media/2009/may/18/mps-expenses-how-scoop -came-light?guni=Article:in%20body%20link.

Brynjolfsson, E. and McAfee, A. (2011) *Race against the machine: How the digital revolution is accelerating innovation, driving productivity, and irreversibly transforming employment and the economy.* Lexington, MA: Digital Frontier Press.

Buncombe, A. (2013) Family of Sunil Tripathi – missing student wrongly linked to Boston marathon bombing – thank well-wishers for messages of support. *The [London] Independent*, 26 April. Available: http://www.independent.co.uk/ news/world/americas/family-of-sunil-tripathi–missing-student-wrongly-linked -to-boston-marathon-bombing–thank-wellwishers-for-messages-of-support-85 86850.html.

Burrell, I. (2012) The Twitter 100: No 1 to 10. *The [London] Independent*, 1 March 2012. Available: http://www.independent.co.uk/news/people/news/ the-twitter-100-no-1-to-10-7466795.html.

Burton, S. and Van Der Hoven, A. (2007) Universal access in South Africa: Broadening communication or building infrastructure? in Nightingale, V. and Dwyer, T. (eds) *New media worlds:challenges for convergence.* Melbourne: Oxford University Press, pp. 231–245.

Butler, J. (1990) *Gender trouble.* New York: Routledge.

Caldas-Coulthard, C.R. (2005) Personal web pages and the semiotic construction of academic identities, in Caldas-Coulthard, C.R. and Toolan, M. (eds) *The writer's craft, the culture's technology.* London & New York: Rodopi, pp. 23–46.

Cardoso, G., Liang, G. and Lapa, T. (2013) Cross-national comparative perspectives from The World Internet Project, in Dutton, W.H. (ed.) *The Oxford handbook of Internet studies.* Oxford: Oxford University Press, pp. 216–236.

Carr, N. (2008) Is Google making us stupid? *The Atlantic*, July/August.

Carr, N. (2010) *The shallows: How the Internet is changing the way we think, read and remember.* London: Atlantic Books.

Castells, M. (2010a) *The rise of the network society.* Second edition with a new preface. Oxford: Wiley-Blackwell.

Castells, M. (2010b) *The power of identity*. Second edition with a new preface. Malden, MA: Wiley-Blackwell.

Castells, M., Caraca, J. and Cardoso, G. (2012) The cultures of the economic crisis: An introduction, in Castells, M., Caraca, J. and Cardoso, G. (eds) *Aftermath: The cultures of the economic crisis*. Oxford: Oxford University Press, pp. 1–14.

Cavallo, G. and Chartier, R. (1991) Introduction, in Cavallo, G. and Chartier, R. (eds) *A history of reading in the west* (translated by Lydia G. Cochrane). Cambridge and Oxford: Polity Press and Blackwell Publishing, pp. 1–36.

Cavanagh, A. (2007) *Sociology in the age of the Internet*. Maidenhead: Open University Press.

Cere, R. (2006) The poor's banker: Real or 'virtual' help? The Internet, NGOs and gendered poverty, in Sarikakis, K. and Thussu, D.K. (eds) *Ideologies of the Internet*. Cresskill, NJ: Hampton Press, pp. 283–298.

Chadwick, A. (2006) *Internet politics: States, citizens, and new communication technologies*. New York: Oxford University Press.

Chang, H.-J. (2011) *23 things they don't tell you about capitalism*. London: Penguin.

Charney, D. (1994) The effect of hypertext on processes of reading and writing, in Selfe, C.L. and Hilligoss, S. (eds) *Literacy and computers: The complications of teaching and learning with technology*. New York: Modern Language Association, pp. 238–263.

Cheong, P.H. and Gong, J. (2010) Cyber vigilantism, transmedia collective intelligence, and civic participation. *Chinese Journal of Communication*, 3(4): 471–487.

Chiong, C., Ree, J., Takeuchi, L. and Erickson, I. (2012) *Print books vs. e-books: Comparing parent-child co-reading on print, basic, and enhanced e-book platforms*. New York: The Joan Ganz Cooney Center. Available: http://www.joanganz cooneycenter.org/wp-content/uploads/2012/07/jgcc_ebooks_quickreport.pdf.

Christensen, C. (2011) Twitter revolutions? Addressing social media and dissent. *The Communication Review*, 14(3): 155–157.

CIBER (2007) *Information behaviour of the researcher of the future. Work package V. Trends in scholarly information behaviour: Technology trends*. By Barrie Gunter. Available: http://www.jisc.ac.uk/media/documents/programmes/rep pres/ggworkpackagev.pdf.

CNN Money (2007) *Fortune 500*. Available: http://money.cnn.com/magazines/ fortune/fortune500/2007/industries/.

CNN Money (2011) *Global 500*. Available: http://money.cnn.com/magazines/ fortune/global500/2011/full_list/.

CNN Money (2013) *Fortune 500*. Available: http://money.cnn.com/magazines/ fortune/fortune500/2013/full_list/index.html?industry.

Comunello, F. and Anzera, G. (2012) Will the revolution be tweeted? A conceptual framework for understanding the social media and the Arab Spring. *Islam and Christian-Muslim Relations*, 23(4): 453–470.

Couldry, N. (2012) *Media, society, world: Social theory and digital media practice*. Cambridge, UK, Malden, MA: Polity Press.

Cox, R.J. (1992) *Vandals in the stacks: A response to Nicholson Baker's assault on libraries*. Westport, CT: Greenwood Press.

Curran, J. (2012) Rethinking Internet history, in Curran, J., Fenton, N. and Freedman, D. (eds) *Misunderstanding the Internet*. Abingdon and New York: Routledge, pp. 34–65.

Dahlberg, L. (2007a) Rethinking the fragmentation of the cyberpublic: From consensus to contestation. *New Media & Society*, 9(5): 827–847.

Dahlberg, L. (2007b) The Internet and discursive exclusion: From deliberative to agonistic public sphere theory, in Dahlberg, L. and Siapera, E. (eds) *Radical democracy and the Internet: Interrogating theory and practice*. Basingstoke and New York: Palgrave Macmillan, pp. 128–147.

Dahlgren, P. (2005) The Internet, public spheres, and political communication: Dispersion and deliberation. *Political Communication*, 22(2): 147–162.

Danaher, G., Schirato, T. and Webb, J. (2002) *Understanding Foucault*. Los Angeles, London, New Delhi, Singapore and Washington, DC: Sage.

Darling, A. (2011) *Back from the brink: 1,000 days at number 11*. London: Atlantic Books.

Davies, N. (2008) *Flat earth news*. London: Chatto & Windus.

Demos (2008) *Voter turnout by income, 2008 US presidential election*. New York. Available: http://www.demos.org/data-byte/voter-turnout-income-2008 -us-presidential-election.

DeNinno, N. (2013) 8 photos of the manhunt for Boston bombing suspect Dzhokar Tsarnaev, 19 April. Available: http://www.ibtimes.com/8-photos -manhunt-boston-bombing-suspect-dzhokar-tsarnaev-1204727.

Department for Culture, Media and Sport (DCMS) (UK) (1998) *Creative industries mapping document*. London: DCMS.

Department for Culture, Media and Sport (DCMS) (UK) (2001) *Creative industries mapping document*. London: DCMS.

Department for Culture, Media and Sport (DCMS): Evidence and Analysis Unit (UK) (2007) *The creative economy programme: A summary of projects commissioned in 2006/7*. London: DCMS.

Department for Culture, Media and Sport (DCMS): Evidence and Analysis Unit (UK) (2008) *Creative Britain: New talents for the new economy*. London: DCMS.

Derrida, J. (1996) *Archive fever: A Freudian impression*. Chicago and London: University of Chicago Press.

DeWoskin, R. (2012) East meets tweet. *Vanity Fair*, 17 February. Available: http:// www.vanityfair.com/culture/2012/02/weibo-china-twitter-chinese-microblogg ing-tom-cruise-201202.

Dibbell, J. (2001/1993) A rape in cyberspace: Or how an evil clown, a Haitian trickster spirit, two wizards, and a cast of dozens turned a database into a society, in Trend, D. (ed.) *Reading digital culture*. Malden, MA, Oxford Carlton, Australia: Blackwell Publishing, pp. 199–213.

Dibbell, J. (2001) The paper chase. *Village Voice Literary Supplement*, April 2001.

Doueihi, M. (2011) *Digital cultures*. Cambridge, MA: Harvard University Press.

Downey, J. (2007) Participation and/or deliberation? The Internet as a tool for achieving radical democratic aims, in Dahlberg, L. and Siapera, E. (eds) *Radical democracy and the Internet: Interrogating theory and practice*. Basingstoke and New York: Palgrave Macmillan, pp. 108–127.

Downey, J. and Fenton, N. (2003) New media, counter publicity and the public sphere. *New Media & Society*, 5(2): 185–202.

Downey, T. (2010) China's cyberposse. *New York Times*, 3 March. Available: http:// www.nytimes.com/2010/03/07/magazine/07Human-t.html?pagewanted= all&_r=0.

Drucker, P. (1992/1968) *The age of discontinuity: Guidelines to our changing society*. New Brunswick, NJ: Transaction.

Duguid, P. (2007) Inheritance and loss? A brief survey of Google Books. *First Monday*, 12(8). Available: http://firstmonday.org/htbin/cgiwrap/bin/ojs/index.php/fm/article/view/1972/1847.

Dyer-Witheford (2007) Hegemony or multitude? Two versions of radical democracy for the Net, in Dahlberg, L. and Siapera, E. (eds) *Radical democracy and the Internet: Interrogating theory and practice*. Basingstoke and New York: Palgrave Macmillan, pp. 191–206.

Eco, U. (2006) *On literature*. London: Vintage.

Elberse, A. (2008) Should you invest in the long tail? *Harvard Business Review*, July–August: 1–11.

Elliott, L. and Atkinson, D. (2007) *Fantasy island: Waking up to the incredible economic, political and social illusions of the Blair legacy*. London: Constable.

Ellul, J. (1964/1954) *The technological society*. New York: Vintage Books.

EON: Enhanced Online News (2010) DontDateHimGirl.com set to remove postings of alleged cheaters from site in 30 days, 19 July. Available: http://www.enhancedonlinenews.com/portal/site/eon/permalink/?ndmViewId=news_view&newsId=20100719006160&newsLang=en.

Fallows, D. (2005) Search engine users. *Pew Internet and American Life Project*. Available: http://www.pewinternet.org/Reports/2005/Search-Engine-Users.aspx

Fanon, F. (1952/2008) *Black skins, white masks* (translated by Richard Philcox). New York: Grove Press.

Faris, D. (2013) *Dissent and revolution in a digital age: Social media, blogging and activism in Egypt*. London and New York: I.B. Tauris.

Fenrich, W.J. (1996) Common law protection of individuals' rights in personal information. *Fordham Law Review*, 65: 951–1003.

Fenton, N. (2012) The Internet and social networking, in Curran, J., Fenton, N. and Freedman, D. (eds) *Misunderstanding the Internet*. Abingdon and New York: Routledge, pp. 123–148.

File, T. (2013) *The diversifying electorate – voting rates by race and Hispanic origin in 2012 (and other recent elections)*. Current population survey reports, P20-569. Washington, DC: United States Census Bureau. Available: http://www.census.gov/prod/2013pubs/p20-568.pdf.

Filipovic, J. (2013) 'Revenge porn' is about degrading women sexually and professionally. *The Guardian*, 28 January. Available: http://www.theguardian.com/commentisfree/2013/jan/28/revenge-porn-degrades-women.

Flew, T. (2007) *Understanding global media*. Basingstoke and New York: Palgrave Macmillan.

Flew, T. (2012) *The creative industries: Culture and policy*. London, Thousand Oaks, New Delhi and Singapore: Sage Publications.

Flood, A. (2012) American digital public library promised for 2013. *The Guardian*, 5 April. Available: http://www.theguardian.com/books/2012/apr/05/american-digital-public-library-promise.

Ford, M. (2009) *The lights in the tunnel: Automation, accelerating technology and the economy of the future*. USA: Acculant Publishing.

Foucault, M. (1991) Governmentality, in Burchell, G. (ed.) *The Foucault effect*. Chicago: Chicago University Press, pp. 87–104.

Fowler, J. (2014) ILO warns of jobless recovery. *Agence France-Presse*, 21 January 2014. Available: http://www.abs-cbnnews.com/business/01/21/14/ilo-warns -jobless-recovery.

Fraser, N. (1992) Rethinking the public sphere: A contribution to the critique of actually existing democracy, in Calhoun, C. (ed.) *Habermas and the public sphere*. Cambridge, MA and London: MIT Press, pp. 109–142.

Freedman, D. (2012) Web 2.0 and the death of the blockbuster economy, in Curran, J., Fenton, N. and Freedman, D. (eds) *Misunderstanding the Internet*. Abingdon and New York: Routledge, pp. 69–94.

Friedman, T. (2006) *The world is flat: The globalized world in the twenty-first century*. London and New York: Penguin.

Fuchs, C. (2008) *The Internet and society: Social theory and the information age*. London and New York: Routledge.

Fuchs, V. (1980) *Economic growth and the rise of service sector employment*. Working paper no. 486. Cambridge, MA: National Bureau of Economic Research.

G20 Meltdown in the City (2009): http://www.g-20meltdown.org/ (website no longer available).

Galperin, E. (2012) How to remove your Google search history before Google's new privacy policy takes effect. *Electronic Frontier Foundation*, 21 February. Available: https://www.eff.org/deeplinks/2012/02/how-remove -your-google-search-history-googles-new-privacy-policy-takes-effect.

Garside, J. (2012) Google's privacy policy 'too vague'. *The Guardian*, 8 March. Available: http://www.theguardian.com/technology/2012/mar/08/ google-privacy-policy-too-vague.

Garside, J. (2013a) Google and privacy: European data regulators round on search giant. *The Guardian*, 20 June. Available: http://www.theguardian.com/ technology/2013/jun/20/google-privacy-european-regulators.

Garside, J. (2013b) Nasdaq crash triggers fear of data meltdown. *The Guardian*, 23 August. Available: http://www.theguardian.com/technology/2013/aug/23/ nasdaq-crash-data.

Gates III, W.H. (with Myrhvold, N. and Rinearson, P.) (1995) *The road ahead*. London: Penguin.

Gellman, B. and Poitras, L. (2013) U.S., British intelligence mining data from nine U.S. Internet companies in broad secret program. *The Washington Post*, 7 June. Available: http://www.washingtonpost.com/investigations/us -intelligence-mining-data-from-nine-us-internet-companies-in-broad-secret -program/2013/06/06/3a0c0da8-cebf-11e2-8845-d970ccb04497_story.html.

Gibbs, S. (2013) Icann chief: Shift away from US 'is the way forward'. *The Guardian*, 21 November. Available: http://www.theguardian.com/technology/ 2013/nov/21/icann-internet-governance-solution-us-nsa-brazil-argentina.

Giddens, A. (1991) *Modernity and self-identity: Self and society in the late modern age*. Cambridge: Polity Press.

Gilliland-Swetland, A.J. (2000) *Enduring paradigm, new opportunities: The value of the archival perspective in the digital environment*. Washington, DC: Council on Library and Information Resources. Available: http://www.clir.org/pubs/ reports/pub89/pub89.pdf.

Gillmor, D. (2010) *Mediactive* (self-published).

Gleick, J. (2011) How Google dominates us. *New York Review of Books*, 18 August.

Glenton, J. (2013) Soldier worship blinds Britain to the grim reality of war. *The Guardian*, 8 November. Available: http://www.theguardian.com/commentisfree/2013/nov/08/soldier-worship-royal-marine-murder-afghan.

Godin, S. (2008) *Tribes: We need you to lead us*. London: Piatkus.

Golumbia, D. (2013) *Cyberlibertarianism: The extremist foundations of 'digital freedom'*. Talk delivered at Clemson University, 5 September. Available: http://www.uncomputing.org/wp-content/uploads/2013/09/cyberlibertarianism-extremist-foundations-sep2013.pdf.

Gore, A. (1994) *Remarks delivered at the meeting of the International Telecommunications Union*, Buenos Aires, 21 March.

Green, J. (2006) Blind date. *Miami New Times*, 14 September. Available: http://www.miaminewtimes.com/2006-09-14/news/blind-date/.

Greenfield, S. (2006) *Proceedings of the [British] House of Lords*, 20 April. Available: http://www.publications.parliament.uk/pa/ld200506/ldhansrd/vo060420/text/60420-18.htm.

Gregory, A. (2009) Man dies at G20 protests. *Daily Mirror*, 2 April 2009. Available: http://www.mirror.co.uk/news/uk-news/man-dies-in-g20-protests-385819.

Growth Intelligence/National Institute of Economic and Social Research (NIESR) (2013) *Measuring the UK's digital economy with big data*. Available: http://niesr.ac.uk/sites/default/files/publications/SI024_GI_NIESR_Google_Report12.pdf.

Habermas, J. (1991/1962) *The structural transformation of the public sphere: An inquiry into a category of bourgeois society*. Cambridge, MA: MIT Press.

Habermas, J. (1996) *Between facts and norms* (translated by W. Rehg). Cambridge, MA: MIT Press.

Habermas, J. (1997/1989) The public sphere, in Goodin, R.E. and Pettit, P. (eds) *Contemporary political philosophy. An anthology*. Oxford and Malden, MA: Blackwell, pp. 105–108.

Habermas, J. (2006) *Towards a United States of Europe*. Acceptance speech for the Bruno Kreisky Prize, 9 March. Available: http://www.signandsight.com/features/676.html.

Hackett, R.A. (1986) For a socialist perspective on the news media. *Studies in Political Economy*, 19(Spring): 141–156.

Halavais, A.M.C. (2009) *Search engine society*. London: Polity Press.

Halliday, J. (2011) David Cameron considers banning suspected rioters from social media. *The Guardian*, 11 August. Available: http://www.theguardian.com/media/2011/aug/11/david-cameron-rioters-social-media.

Halpern, S. (2012) Are hackers heroes? *New York Review of Books*, 27 September.

Hamilton, T. (2003) Getting to know you: Opening a bank account gives regulatory datacorp a window on your life. *Toronto Star*, 18 June.

Hampton, K., Sessions Goulet, L., Rainie, L. and Purcell, K. (2011) *Social networking sites and our lives: How people's trust, personal relationships, and civic and political involvement are connected to their use of social networking and other technologies*. Washington, DC: Pew Internet & American Life Project. Available: http://www.pewinternet.org/~/media//Files/Reports/2011/PIP%20-%20Social%20networking%20sites%20and%20our%20lives.pdf.

Hands, J. (2011) *@ is for activism: Dissent, resistance and rebellion in a digital culture*. London and New York: Pluto Press.

Haraway, D. (1985) Manifesto for cyborgs: Science, technology and socialist feminism in the 1980s. *Socialist Review*, 80: 65–108.

Hardt, M. and Negri, A. (2000) *Empire.* Cambridge, MA, London: Harvard University Press.

Hardt, M. and Negri, A. (2006) *Multitude: War and democracy in the age of empire.* London: Penguin Books.

Harford, T. (2012) *High frequency trading and the $440m mistake.* London: BBC, 10 August. Available: http://www.bbc.co.uk/news/magazine-19214294.

Harvey, D. (1990) *The condition of postmodernity.* Malden, MA, Oxford: Blackwell Publishers.

Hayes, C. (2013) Before PRISM there was Total Information Awareness. *MSNBC,* 12 September 2013. Available: http://www.msnbc.com/all-in/prism-there-was -total-information-awar.

Head, A.J. (2007) Beyond Google: How do students conduct academic research? *First Monday,* 12(8). Available: http://www.uic.edu/htbin/cgiwrap/bin/ojs/ index.php/fm/article/view/1998/1873.

Head, A.J. and Eisenberg, M.B. (2009) Lessons learned: How college students seek information in the digital age. Project Information Literacy progress report. University of Washington. Available: http://projectinfolit.org/pdfs/PIL _Fall2009_Year1Report_12_2009.pdf.

Heidegger, M. (2003/1977) The question concerning technology, in Scharff, R.C. and Dusek, V. (eds) *Philosophy of technology: The technological condition. An anthology.* Malden, MA, Oxford, Melbourne and Berlin: Blackwell Publishing, pp. 252–264.

Heilbroner, R.L. (2003/1967) Do machines make history? in Scharff, R.C. and Dusek, V. (eds) *Philosophy of technology: The technological condition. An anthology.* Malden, MA, Oxford, Melbourne and Berlin: Blackwell Publishing, pp. 398–404.

Hesmondhalgh, D. (2007) *The cultural industries.* Second edition. Los Angeles, London, New Delhi and Singapore: Sage Publications.

Hewitt, D. (2012) Weibo brings change to China. *BBC News,* 31 July. Available: http://www.bbc.co.uk/news/magazine-18887804.

Hill, A. (2009) *Re-imagining the war on terror: Seeing, waiting, travelling.* Basingstoke and New York: Palgrave Macmillan.

Hill, S. (2009) The world wide webbed: The Obama campaign's masterful use of the Internet. *Social Europe Journal,* 4(2): 9–15.

Hill, S. (2013) *Digital revolutions: Activism in the digital age.* Oxford: New International Publications.

Hirschorn, M. (2008) Only connect. *Atlantic Monthly,* 1 May. Available: http:// www.theatlantic.com/magazine/archive/2008/05/only-connect/306766/.

Holcomb, J. (2013) *How the Lehman Bros. crisis impacted the 2008 presidential race.* Washington, DC: Pew Research Center. Available: http://www.pew research.org/fact-tank/2013/09/19/how-the-lehman-bros-crisis-impacted-the -2008-presidential-race/.

Hopkins, N. (2013) UK gathering secret intelligence via covert NSA operation. *The Guardian,* 7 June 2013. Available: http://www.theguardian.com/technology/ 2013/jun/07/uk-gathering-secret-intelligence-nsa-prism?CMP=twt_gu.

HPC Wire (2010) *Algorithmic terrorism on Wall Street,* 5 August. Available: http:// www.hpcwire.com/hpcwire/2010-08-05/algorithmic_terrorism_on_wall_street .html.

Huyer, S. (2006) E-quality or e-poverty? Gender and international policy on ICT for development, in Sarikakis, K. and Thussu, D.K. (eds) *Ideologies of the Internet.* Cresskill, NJ: Hampton Press, pp. 213–243.

Indergaard, M. (2013) Beyond the bubbles: Creative New York in boom, bust and the long run. *Cities*, 33: 43–50.

Indymedia London (2009) *Witness statement about G20 death*, 2 April 2009. Available: http://london.indymedia.org/videos/1023.

International Commission for the Study of Communication Problems (1980) *Many voices, one world.* London, New York and Paris: UNESCO. Available: http:// unesdoc.unesco.org/images/0004/000400/040066eb.pdf.

International Telecommunication Union (ITU) (2013) *Measuring the information society 2013.* Geneva. Available: http://www.itu.int/en/ITU-D/Statistics/ Documents/publications/mis2013/MIS2013_without_Annex_4.pdf.

Internet Governance Forum (IGF) (2011) *What is the Internet Governance Forum? background note. Nairobi*, 27–30 September 2011. Available via website: http:// www.intgovforum.org/cms/aboutigf.

Jacques, M. (2009) *When China rules the world: The rise of the Middle Kingdom and the end of the western world.* London: Allen Lane.

James, R. (2010) An assessment of the legibility of Google Books. *Journal of Access Services*, 7(4): 223–228. Available: http://scholarspace.manoa.hawaii .edu/bitstream/10125/15358/1/google1.pdf.

Jarvis, J. (2009) *What would Google do?* New York: Collins Business.

Jeanneney, J.-N. (2007) *Google and the myth of universal knowledge*, Chicago and London: Chicago University Press.

Johnson, B. (2009) US relinquishes control of the Internet. *The Guardian*, 30 September. Available: http://www.theguardian.com/technology/2009/sep/ 30/icann-agreement-us.

Joint Information Systems Committee (JISC) (2007) *JISC digitization conference. Cardiff*, 20–21 July 2007. Bristol: JISC.

Jones, E. (2010) Google Books as a general research collection. *Library Resources & Technical Services*, 54(2): 77–89.

Jordan, T. (2007) Online direct action: Hacktivism and radical democracy, in Dahlberg, L. and Siapera, E. (eds) *Radical democracy and the Internet: Interrogating theory and practice.* Basingstoke and New York: Palgrave Macmillan, pp. 73–88.

Jordan, T. and Taylor, P.A. (2004) *Hacktivism and cyberwars: Rebels with a cause?* London and New York: Routledge.

Juris, J.S. (2008) *Networking futures: The movements against corporate globalization.* Durham and London: Duke University Press.

Kahneman, D. (2011) *Thinking fast and slow.* Penguin: London and New York.

Kaid, L.L. (2000) Ethics and political advertising, in Denton Jr, R.E. (ed.) *Political communication ethics: An oxymoron?* Westport, CT: Praeger, pp. 147–178.

Keen, A. (2007) *The cult of the amateur: How today's Internet is killing our culture and assaulting our economy.* London and Boston: Nicholas Brealey Publishing.

Keen, A. (2012) *Digital vertigo: How today's online social revolution is dividing, diminishing, and disorienting us.* London: Constable.

Kelly, H. (2013) Police-scanner sites see surge in traffic after Boston. *CNN*, 24 April. Available: http://edition.cnn.com/2013/04/23/tech/web/police-scanners-online-boston/.

Kelly, K. (1999) The roaring zeros. *Wired*, 7(9), September 1999. Available: http://www.wired.com/wired/archive/7.09/zeros.html?pg=1&topic=&topic_set=.

Kelly, K. (2009) The new socialism: Global collectivist society is coming online. *Wired*, 17(6), 22 May. Available: http://www.wired.com/culture/culturereviews/magazine/17-06/nep_newsocialism?currentPage=all.

Kelly, K. (2010) *What technology wants*. London: Penguin Books.

Khondker, H.H. (2011) Role of the new media in the Arab Spring. *Globalizations*, 8(5): 675–679.

Klein, N. (2000) *No logo*. London: Flamingo.

Knight, L. (2013) *A dark magic: The rise of the robot traders*. London: BBC. 7 July. Available: http://www.bbc.co.uk/news/business-23095938.

Kubey, R. (ed.) (1997) *Media literacy in the information age. Information and behaviour*. London and New Brunswick: Transaction Publishers.

Kuittinen, T. (2013) Is the bizarre timing of Blackberry's death tied to Nokia's panic sale to Microsoft? *Forbes*, 20 September. Available: http://www.forbes.com/sites/terokuittinen/2013/09/20/is-the-bizarre-timing-of-blackberrys-death-tied-to-nokias-panic-sale-to-microsoft/.

Kumar, N. (2012) Ohio's car industry voters help to steer Obama back to the White House. *The [London] Independent*, 7 November. Available: http://www.independent.co.uk/news/world/americas/ohios-car-industry-voters-help-to-steer-obama-back-to-the-white-house-8290012.html.

Kumar, N. (2013) His arm in a cast, his accent changed, Dzhokhar Tsarnaev pleads not guilty to Boston marathon bombing charges. *The [London] Independent*, 10 July. Available: http://www.independent.co.uk/news/world/americas/his-arm-in-a-cast-his-accent-changed-dzhokhar-tsarnaev-pleads-not-guilty-to-boston-marathon-bombing-charges-8700744.html.

Kymlicka, W. (2002) *Contemporary political philosophy: An introduction*. Second edition. Oxford and New York: Oxford University Press.

Landow, G.P. (1992) *Hypertext: The convergence of contemporary critical theory and technology*. Baltimore: Johns Hopkins University Press.

Lanier, J. (2011) *You are not a gadget: A manifesto*. London: Penguin.

Lanier, J. (2013) *Who owns the future?* London and New York: Allen Lane.

Lavabit (2013) *Lavabit* home page. Available: http://lavabit.com/.

Leadbetter, C. (2008) *We-think*. London: Profile Books.

Leveson, Lord Justice (2012) *An enquiry into the culture, practices and ethics of the press: Report HC780*. London: TSO (Stationery Office). Available: http://www.official-documents.gov.uk/document/hc1213/hc07/0780/0780.asp.

Lewis, P. (2009) Ian Tomlinson death: Guardian video reveals police attack on man who died at G20 protest. *The Guardian*, 7 April. Available: http://www.theguardian.com/uk/2009/apr/07/ian-tomlinson-g20-death-video.

Lewis, P., Laville, S. and Vidal, J. (2009) Fears police tactics at G20 protests will lead to violence. *The Guardian*, 27 March 2009. Available: http://www.theguardian.com/uk/2009/mar/27/g20-protest.

Leys, N. (2013) After Boston name blame, chief admits 'last and accurate better than first and mistaken'. *The Australian*, 6 May. Available: http://www.theaustralian.com.au/media/after-boston-name-blame-chief

-admits-last-and-accurate-better-than-first-and-mistaken/story-e6frg996-122 6635627236#.

Library of Congress (2013) *Technical notes: By type of material.* Available: http://memory.loc.gov/ammem/dli2/html/document.html.

Lister, M., Dovey, J., Giddings, S. and Kelly, K. (2003) *New media. A critical introduction.* London and New York: Routledge.

Livingston, S. (1997) Clarifying the CNN effect: An examination of media effects according to type of military intervention. *The Joan Shorenstein Center on the Press, Politics and Public Policy, John F. Kennedy School of Government, Harvard University.* Research Paper R-18, June 1997. Available: http://www.genocide-watch.org/images/1997ClarifyingtheCNNEffect-Livingston.pdf.

Livingstone, S. (2004) What is media literacy? *Intermedia*, 32(3): 18–20.

Lovink, G. (2008) *Zero comments: Blogging and critical Internet culture.* London and New York: Routledge.

Lovink, G. (2011) *Networks without a cause: A critique of social media.* Cambridge, UK, Malden, MA: Polity Press.

Lull, J. (2000) *Media, communication, culture: A global approach.* Second edition. Malden, MA, Cambridge, UK: Polity Press.

Machlup, F. (1962) *The production and distribution of knowledge in the United States.* Princeton, NJ: Princeton University Press.

Mack, A.M. (2004) How the Internet is changing politics. *AdWeek*, 26 January. Available: http://www.adweek.com/news/advertising/how-internet-changing-politics-69898.

MacKenzie, D. and Spears, T. (2012) *'The formula that killed Wall Street?' The Gaussian copula and the material cultures of modelling.* Available: http://www.sps.ed.ac.uk/__data/assets/pdf_file/0003/84243/Gaussian14.pdf.

Madden, M. (2009) The 'death panels' are already here. *Salon*, 11 August. Available: http://www.salon.com/2009/08/11/denial_of_care/.

Madrigal, A.C. (2013) Boston bombing: The anatomy of a misinformation disaster. *The Atlantic*, 19 April. Available: http://m.theatlantic.com/technology/archive/2013/04/it-wasnt-sunil-tripathi-the-anatomy-of-a-misinformation-disaster/275155/.

Maffesoli, M. (1996) *The time of the tribes: The decline of individualism in mass society.* London: Sage.

Maier-Rabler, U. (2006) Reconceptualizing e-policy: From bridging the digital divide to closing the knowledge gap, in Sarikakis, K. and Thussu, D.K. (eds) *Ideologies of the Internet.* Cresskill, NJ: Hampton Press, pp. 195–212.

Manguel, A. (2008) *The library at night.* New Haven and London: Yale University Press.

Manovich, L. (2007) Database as symbolic form, in Vesna, V. (ed.) *Database aesthetics: Art in the age of information overflow.* Minneapolis: University of Minnesota Press, pp. 39–60.

Mason, P. (2012) *Why it's kicking off everywhere: The new global revolutions.* London and New York: Verso.

Mayer-Schönberger, V. (2009) *Delete.* Princeton, New Jersey & Oxford: Princeton University Press.

McChesney, R.W. (2013) *Digital disconnect: How capitalism is turning the Internet against democracy.* New York: The New Press.

McGarry, J. and O'Leary, B. (1995) *Explaining Northern Ireland: Broken images*. Oxford: Blackwell.

McGirt, E. (2009) How Chris Hughes helped launch Facebook and the Barack Obama campaign. *Fast Company*, April 2009. Available: http://www.fast company.com/1207594/how-chris-hughes-helped-launch-facebook-and -barack-obama-campaign.

McLuhan, M. (1964) *Understanding media*. London and New York: Routledge.

McNeely, I.F. and Wolverton, L. (2008) *Reinventing knowledge: From Alexandria to the Internet*. London and New York: W.W. Norton.

Mejias, U.A. (2012) Liberation technology and the Arab Spring: From utopia to atopia and beyond. *The Fibreculture Journal*, 20. Available: http://twenty.fibre culturejournal.org/2012/06/20/fcj-147-liberation-technology-and-the-arab -spring-from-utopia-to-atopia-and-beyond/.

Miall, D.S. and Dobson, T. (2001) Reading hypertext and the experience of literature. *Journal of Digital Information*, 2(1). Available: http://journals.tdl.org/jodi/ index.php/jodi/article/view/35/37.

Michael, D.C. and Zhou, Y. (2010) *China's digital generations 2.0: Digital media and commerce go mainstream*. Beijing: BCG (The Boston Consulting Group). Available: http://www.bcg.com/documents/file45572.pdf.

Michaels, S. (2013) Daft Punk's Random Access Memories is Amazon's best-selling vinyl ever. *The Guardian*, 18 September. Available: http://www.theguardian .com/music/2013/sep/18/daft-punk-random-access-memories-amazon.

Miller, V. (2011) *Understanding digital culture*. Los Angeles, London, New Delhi and Singapore: Sage Publications.

Ministry of Justice (UK) (2012) *Freedom of information guidance. Exemptions guidance. Section 43: Commercial interests*. Available: http://www.justice.gov.uk/ downloads/information-access-rights/foi/foi-s43-exemptions.pdf.

MIT Technology Review (2013) 50 disruptive companies: Ushahidi. Available: http://www2.technologyreview.com/tr50/ushahidi/.

Montgomery, L. (2010) *China's creative industries: Copyright, social network markets and the business of culture in a digital age*. Cheltenham, UK and Northampton, MA, USA: Edward Elgar.

Morozov, E. (2013) *To save here, click everything: The folly of technological solutionism*. London: Allen Lane.

Morris, S. (2013) Marine court martial hinged on video of disturbing scenes from Afghan war. *The Guardian*, 8 November. Available: http://www.theguardian. com/uk-news/2013/nov/08/royal-marines-court-martial-video-afghanistan -war.

Mosco, V. (2005) *The digital sublime: Myth, power and cyberspace*. Cambridge, MA: MIT Press.

Mouffe, C. (2005) *On the political*. London: Routledge.

Moulthrop, S. (1995) *Victory garden* [sampler]. Watertown, MA: Eastgate Systems. Available: http://www.eastgate.com/VG/VGStart.html.

Naisbitt, J. (1982) *Megatrends*. New York: Warner Books.

National Library of Australia (NLA) (2013) Trove: Digitized newspapers and more. Available: http://trove.nla.gov.au/newspaper.

Neilson, B. and Rossiter, N. (2005) From precarity to precariousness and back again: Labour, life and unstable networks. *The Fibreculture Journal*, 5. Available: http://five.fibreculturejournal.org/fcj-022-from-precarity -to-precariousness-and-back-again-labour-life-and-unstable-networks/.

NHS (UK) (2013) *Summary care record*. Available: http://www.nhscarerecords .nhs.uk/.

Norris, P. (2000) Global governance and cosmopolitan citizens, in Nye, J.S. and Donahue, J.D. (eds) *Governance in a globalizing world*. Washington, DC: Brookings Institution, pp. 155–177.

OECD (1994) *Employment/unemployment study: Policy report*. Paris: OECD.

O'Neil, M. (2009) *Cyberchiefs: Autonomy and authority in online tribes*. London: Pluto Press.

Ong, W.J. (1982) *Orality and literacy: The technologizing of the word*. London and New York: Methuen.

Osnos, E. (2013) Letter from China: Solzhenitsyn, Yao Chen, and Chinese reform. *The New Yorker*, 8 January. Available: http://www.newyorker.com/online/blogs/ evanosnos/2013/01/solzhenitsyn-yao-chen-and-battle-over-chinese-reform .html.

Palfrey, J. and Gasser, U. (2008) *Born digital*. New York: Basic Books.

Papacharissi, Z. (2010) *A private sphere: Democracy in a digital age*. Cambridge: Polity.

Parekh, B. (2008) *A new politics of identity: Political principles for an interdependent world*. Basingstoke and New York: Palgrave Macmillan.

Pasquali, F. (2014) For an archaeology of online participatory literary writing: Hypertext and hyperfiction. *CM Communication Management Quarterly*, 30: 35–54. Available: http://www.cost-transforming-audiences.eu/system/files/ pub/CM30-SE-Web.pdf.

Paterson, C. (2006) *News agency dominance in international news on the Internet*. Papers in international and global communication, number 01/06. Centre for International Communications Research, Leeds University.

Paul, C. (2007) The database as system and cultural form: Anatomies of cultural narratives, in Vesna, V. (ed.) *Database aesthetics: Art in the age of information overflow*. Minneapolis: University of Minnesota Press, pp. 95–109.

Payne, T. (2013) Police officer filmed allegedly punching a student will not face disciplinary action, Met confirms. *The [London] Independent*, 18 December. Available: http://www.independent.co.uk/student/news/police -officer-filmed-allegedly-punching-a-student-will-not-face-disciplinary-action -met-confirms-9012441.html.

PBS Frontline (2010) *The secret online life of Autumn Edows*. Available: http:// www.pbs.org/wgbh/pages/frontline/digitalnation/relationships/identity/ the-secret-online-life-of-autumn-edows.html.

Peev, G. (2010) Google 'revealed address of battered wife refuge on Street View'. *Mail Online*, 29 October. Available: http://www.dailymail.co.uk/news/ article-1324637/Google-Street-View-revealed-address-battered-wife-refuge .html.

Petrucci, A. (1991) Reading to read: A future for reading, in Cavallo, G. and Chartier, R. (eds) *A history of reading in the west* (translated by Lydia G. Cochrane). Cambridge and Oxford: Polity Press and Blackwell Publishing, pp. 345–367.

Pew Research Center (2008) *Inside Obama's sweeping victory*. Washington, DC. Available: http://www.pewresearch.org/2008/11/05/inside-obamas-sweeping -victory/.

Pew Research Center (2013a) *China has more Internet users than any other country*. Washington, DC. Available: http://www.pewresearch.org/fact-tank/2013/12/02/china-has-more-internet-users-than-any-other-country/.

Pew Research Center (2013b) *Republican support for war at all-time low: A decade later, Iraq war divides the public*. Washington, DC. Available: http://www.people-press.org/files/legacy-pdf/3-18-13%20Iraq%20Release.pdf.

Pew Research Center's Project for Excellence in Journalism (2008) *McCain v. Obama on the Web: A study of the presidential candidate web sites* [*sic*]. Washington, DC. Available: http://www.journalism.org/files/legacy/CAMPAIGN_WEB_08_DRAFT_IV_copyedited.pdf.

Plant, S. (1996) On the matrix: Cyberfeminist simulations, in Shields, R. (ed.) *Cultures of the Internet: Virtual spaces, real histories, living bodies*. London: Sage Publications, pp. 170–183.

Polastron, L.X. (2007) *Books on fire: The tumultuous story of the world's great libraries* (translated by Jon E. Graham). London: Thames & Hudson.

Pope, J. (2013) The way ahead: The teaching of hyper-narrative at Bournemouth University. *New Writing: The International Journal for the Practice and Theory of Creative Writing*, 10(2): 206–218.

Porat, M.U. (1977) *The information economy: Definition and measurement*. Washington, DC: Office of Telecommunications (DOC).

Porter, H. (2009) Google is just an amoral menace. *The Observer*, 5 April 2009. Available: http://www.theguardian.com/commentisfree/2009/apr/05/google-internet-piracy.

Poster, M. (1995a) Postmodern virtualities, in Featherstone, M. and Burrows, R. (eds) *Cyberspace cyberbodies cyberpunk: Cultures of technological embodiment*. London: Sage, pp. 79–95.

Poster, M. (1995b) *The second media age*. Cambridge: Polity.

Poster, M. (1996) Databases as discourse: Or, electronic interpellations, in Lyon, D. and Zureik, E. (eds) *Computers, surveillance, and privacy*. Minneapolis & London: University of Minnesota Press, pp. 175–193.

Poster, M. (2001/1997) Cyberdemocracy: The Internet and the public sphere, in Trend, D. (ed.) *Reading digital culture*. Malden, MA, Oxford Carlton, Australia: Blackwell Publishing, pp. 259–271.

Poster, M. (2006) *Information please*. Durham and London: Duke University Press.

Postman, N. (1993) *Technopoly: The surrender of culture to technology*. New York: Vintage Books.

Pratt, A.C. and Hutton, T.A. (2013) Reconceptualising the relationship between the creative economy and the city: Learning from the financial crisis. *Cities*, 33: 86–95.

Qiu, J.L. (2013) Network societies and Internet studies: Rethinking time, space, and class, in Dutton, W.H. (ed.) *The Oxford handbook of Internet studies*. Oxford: Oxford University Press, pp. 109–128.

Rahimi, B. (2011) The agonistic social media: Cyberspace in the formation of dissent and consolidation of state power in postelection Iran. *The Communication Review*, 14(3): 158–178.

Raley, R. (2009) *Tactical media*. Minneapolis and London: University of Minnesota Press.

Ramsey, P. (2013) The search for a civic commons online: An assessment of existing BBC Online policy. *Media, Culture & Society*, 35(7): 864–879.

Rap News Network (2006) *Tower records turn to digital downloads*, 28 June. Available: http://www.rapnews.net/0-202-261614-00.html.

Reddit (2013) Reflections on the recent Boston crisis, 22 April. Available: http://blog.reddit.com/2013/04/reflections-on-recent-boston-crisis.html.

Resnick, P., Zeckhauser, R., Swanson, J. and Lockwood, K. (2006) The value of reputation on eBay: A controlled experiment. *Experimental Economics*, 9(2): 79–101.

Rettberg, J.W. (2008) *Blogging*. Cambridge and Malden, MA: Polity Press.

Rheingold, H. (2002) *Smart mobs: The next social revolution*. Cambridge, MA: Basic Books.

Risen, T. (2013) Study: NSA spying may cost U.S. companies $35 billion. *U.S. News & World Report*, 27 November. Available: http://www.usnews.com/news/articles/2013/11/27/study-nsa-spying-may-cost-us-companies-35-billion.

Ritholtz, B. (2013) How Walmart's low wages cost all Americans, not just its workers. *Huffington Post (Bloomberg View)*, 18 December. Available: http://www.huffingtonpost.com/2013/12/18/walmart_n_4466850.html.

Rogers, S. (2009a) How to crowdsource MPs' expenses. *The Guardian*, 18 June. Available: http://www.theguardian.com/news/datablog/2009/jun/18/mps-expenses-houseofcommons.

Rogers, S. (2009b) MPs' expenses: What you've told us. So far. *The Guardian*, 2 November. Available: http://www.theguardian.com/news/datablog/2009/sep/18/mps-expenses-westminster-data-house-of-commons.

Rojecki, A. (2002) Modernism, state sovereignty and dissent: Media and the new post-cold war movements. *Critical Studies in Media Communication*, 19(2): 152–171.

Rolls, C. and October, A. (2004) *Report on the use of ICTs by the Anti-Privatisation Forum*. Transcriptions of interviews conducted with APF staff members and volunteers, November (in possession of Herman Wasserman).

Rossiter, N. (2006) *Organized networks: Media theory, creative labour, new institutions*. Rotterdam: NAi Publishers.

Roszak, T. (1988) *The cult of information: The folklore of computers and the true art of thinking*. London: Paladin.

Rowlands, I., Nicholas, D., Williams, P., Huntington, P., Fieldhouse, M., Gunter, B., Withey, R., Jamali, H.R., Dobrowolski, T. and Tenopir, C. (2008) The Google generation: The information behavior of the researcher of the future. *Aslib Proceedings: New Information Perspectives*, 60(4): 290–310.

Rowthorn, B. (2001) *UK competitiveness and the knowledge economy*. Cambridge: MIT Institute.

Rushe, D. (2013) Zuckerberg: US government 'blew it' on NSA surveillance. *The Guardian*, 12 September. Available: http://www.theguardian.com/technology/2013/sep/11/yahoo-ceo-mayer-jail-nsa-surveillance.

Said, E.W. (2003) *Orientalism. Reprinted with a new preface*. London: Penguin Books.

Sales, B. (2013) Fifty shades of grey: The new publishing paradigm. *Huffington Post (The Blog)*, 18 April. Available: http://www.huffingtonpost.com/bethany-sales/fifty-shades-of-grey-publishing_b_3109547.html.

Sandel, M.J. (1998) *Liberalism and the limits of justice*. Second edition. Cambridge: Cambridge University Press.

Sargent, M.-R. (ed.) (1999) *Francis Bacon: Selected philosophical works*. Indianapolis: Hackett Publishing Company.

Sarikakis, K. (2006) Mapping the ideologies of Internet policy, in Sarikakis, K. and Thussu, D.K. (eds) *Ideologies of the Internet*. Cresskill, NJ: Hampton Press, pp. 163–178.

Sarikakis, K. and Thussu, D.K. (2006) The Internet as ideology, in Sarikakis, K. and Thussu, D.K. (eds) *Ideologies of the Internet*. Cresskill, NJ: Hampton Press, pp. 1–16.

Savat, D. (2009) Deleuze's objectile: From discipline to modulation, in Poster, M. and Savat, D. (eds) *Deleuze and new technology*. Edinburgh: Edinburgh University Press, pp. 45–62.

Schiller, H.I. (2001/1995) The global information highway: Projects for an ungovernable world, in Trend, D. (ed.) *Reading digital culture*. Malden, MA, Oxford Carlton, Australia: Blackwell Publishing, pp. 159–171.

Seidensticker, B. (2006) *Future hype: The myths of technology change*. San Francisco: Berrett-Koehler Publishers.

Selwyn, N. (2004) Reconsidering political and popular understandings of the digital divide. *New Media & Society*, 6(3): 341–362.

Sen, A. (2007) *Identity and violence: The illusion of destiny*. London: Penguin Books.

Shaxson, N. (2011) *Treasure islands: Tax havens and the men who stole the world*. London: The Bodley Head.

Shirky, C. (2009) *Here comes everybody: The power of organizing without organizations*. London: Penguin.

Shirky, C. (2010) *Cognitive surplus: How technology makes consumers into collaborators*. London and New York: Penguin.

Shirky, C. (2011) The political power of social media. *Foreign Affairs*, 90(1): 28–41. Available: http://www.foreignaffairs.com/articles/67038/clay-shirky/the-political-power-of-social-media.

Shunmugam, V. (2010) Financial markets regulation: The tipping point. *Vox*, 18 May. Available: http://www.voxeu.org/article/dark-pools-menace-rising-opacity-financial-markets.

Siapera, E. (2012) *Understanding new media*. Los Angeles, London, New Delhi, Singapore: Sage Publications.

Smith, D. and Rogers, R. (2009) Office staff warned of confrontation as city braces for mass G20 protests. *The Observer*, 22 March. Available: http://www.theguardian.com/business/2009/mar/22/g20-anti-globalization-protests.

Smith, M.R. and Marx, L. (eds) (1996) *Does technology drive history?* Cambridge, MA: MIT Press.

Smith, Z. (2010) Generation why? *New York Review of Books*, 25 November.

Solanas, V. (2004/1968) *The SCUM Manifesto*. London: Verso.

Solove, D.J. (2004) *The digital person: Technology and privacy in the information age*. New York: New York University Press.

Stalder, F. (2002) Opinion. Privacy is not the antidote to surveillance. *Surveillance & Society*, 1(1): 120–124. Available: http://library.queensu.ca/ojs/index.php/surveillance-and-society/article/view/3397/3360.

Steedman, C. (2001) *Dust*. Manchester and New York: Manchester University Press.

Stevenson, C. (2014) The secret to Hollywood's future? Think big despite high-cost blockbuster flops such as 47 Ronin – on track to lose 'up to £106m'. *The*

[London] Independent, 18 January. Available: http://www.independent.co.uk/arts-entertainment/films/features/the-secret-to-hollywoods-future-think-big-despite-highcost-blockbuster-flops-such-as-47-ronin–on-track-to-lose-up-to-106m-9069528.html.

Stiglitz, J. (2002) *Globalization and its discontents*. London: Penguin Books.

Stirland, S.L. (2008) Propelled by Internet: Barack Obama wins presidency. *Wired*, 4 November. Available: http://www.wired.com/threatlevel/2008/11/propelled-by-in/.

Stone, A. (2004) Essentialism and anti-essentialism in feminist philosophy. *Journal of Moral Philosophy*, 1(2): 135–153.

Stone, A.R. (1995) *The war of desire and technology at the close of the mechanical age*. Cambridge, MA: MIT Press.

Stross, R. (2008) *Planet Google*. New York: Free Press.

Sunstein, C. (2001) *Republic.com 2.0*. Princeton and Oxford: Princeton University Press.

Sunstein, C. (2007) *Republic.com 2.0*. Princeton and Oxford: Princeton University Press.

Surowiecki, J. (2005) *The wisdom of crowds*. London: Abacus.

Talbot, D. (2011) Crisis mapping meets check-in. *MIT Technology Review*, 28 March. Available: http://www.technologyreview.com/news/423440/crisis-mapping-meets-check-in/.

Tapscott, D. and Williams, A.D. (2007) *Wikinomics: How mass collaboration changes everything*. London: Atlantic Books.

The Economist (2009) Beyond voice: Special report on telecoms in emerging markets, 24 September. Available: http://www.economist.com/node/14483848.

The Guardian (2009) MPs' expenses: What you've discovered, 16 December. Available: http://www.theguardian.com/politics/2009/dec/16/mps-expenses-what-we-learned

The Guardian (2013) Google Maps image shows my dead son, says California man, 19 November. Available: http://www.theguardian.com/technology/2013/nov/19/google-maps-dead-son-californian.

The Huffington Post UK (2013) Prism: David Davis suggests William Hague or Theresa May would have known GCHQ used controversial US programme, 8 June 2013. Available: http://www.huffingtonpost.co.uk/2013/06/08/prism-david-davis-theresa-may-william-hague_n_3407339.html.

[The] *Washington Times* (2013) Marathon bomber manhunt blog: Dzhokar Tsarnaev hid from police in Watertown yard, 19 April. Available: http://www.washingtontimes.com/blog/breaking-news/2013/apr/19/latest-updates-manhunt-suspect-no-2-boston-maratho/?page=all.

Thompson, J.B. (1993) The theory of the public sphere. Review article. *Theory, Culture & Society*, 10: 173–189.

Thompson, M. (2013) Europe turns up heat on Google over privacy. *CNN Money*, 19 February. Available: http://money.cnn.com/2013/02/19/technology/google-eu-privacy/index.html.

Thussu, D.K. (2007) Introduction, in Thussu, D.K. (ed.) *Media on the move: Global flow and counter-flow*. London and New York: Routledge, pp. 1–8.

Tiefenbrun, I. (2006) *Journal of the Royal Society for Arts*. August.

Toffler, A. (1970) *Future shock*. New York: Random House.

Toffler, A. (1980) *The third wave*. New York: Bantam Books.

Tomlinson, J. (2003) The agenda of globalization. *New Formations*, 50: 10–21.

Touraine, A. (1969) *La societe post industrielle*. Paris: Denoel.

Touraine, A. (2002) The importance of social movements. *Social Movement Studies*, 1(1): 89–96.

Townsend, R.B. (2007) Google Books: Is it good for history? *American Historical Association*, 45(6). Available: http://www.historians.org/perspectives/issues/2007/0709/0709vie1.cfm.

Travis, A. (2008) Bigger databases increase risks, say watchdogs. *The Guardian*, 29 October 2008. Available: http://www.theguardian.com/technology/2008/oct/29/data-security-breach-civil-liberty.

Turkle, S. (1995) *Life on the screen: Identity on the age of the Internet*. New York: Simon & Schuster.

Turner, F. (2006) *From counterculture to cyberculture: Stewart Brand, the Whole Earth Network, and the rise of digital utopianism*. Chicago and London: The University of Chicago Press.

UNCTAD (United Nations Conference on Trade and Development) (2010) *Creative economy report*. United Nations. Available: http://unctad.org/en/Docs/ditctab20103_en.pdf.

UNCTAD (2012) *Creative economy programme: e-newsletter, no. 18*, June. Available: http://unctad.org/en/PublicationsLibrary/webditctab2012d3_en.pdf.

UNCTAD and the Government of the Republic of Zambia (2011) *Strengthening the creative industries for development in Zambia*. United Nations. Available: http://unctad.org/en/Docs/ditctab20091_en.pdf.

UNDP (2006) *Human development report 2006, beyond scarcity*. New York: Palgrave Macmillan. Available: http://hdr.undp.org/hdr2006/.

United Nations (UN) (1948) *The Universal declaration of human rights*. Available: http://www.un.org/en/documents/udhr/index.shtml.

Unwin, T. (2013) The Internet and development: A critical perspective, in Dutton, W.H. (ed.) *The Oxford handbook of Internet studies*. Oxford: Oxford University Press, pp. 531–554.

Vaidhyanathan, S. (2011) *The googlization of everything (and why we should worry)*. Berkeley and Los Angeles: University of California Press.

Van Dijck, J. (2010) Search engines and the production of academic knowledge. *International Journal of Cultural Studies*, 13(6): 574–592.

Van Loon, J. (2008) *Media technology: Critical perspectives*. Maidenhead: Open University Press/McGraw-Hill Education.

Van Scoyoc, A.M. and Cason, C. (2006) The electronic academic library: Undergraduate research behavior in a library without books. *Portal: Libraries and the Academy*, 6(1): 47–58.

Vise, D.A., with Malseed, M. (2006) *The Google story*. Basingstoke: Pan Books.

Wade, R.H. (2004) Bridging the digital divide: New route to development or new form of dependency? in Avgerou, C., Ciborra, C. and Land, F. (eds) *The social study of information and communication technology: Innovation, actors, and contexts*. Oxford: Oxford University Press, pp. 185–206.

Wang, S.S. (2012) China's Internet lexicon: The symbolic meaning and commodification of *grass mud horse* in the harmonious society. *First Monday*, 17(1–2). Available: http://journals.uic.edu/ojs/index.php/fm/article/view/3758/3134.

Wasserman, H. (2007) Surfing against the tide: The use of new media technologies for social activism in South Africa, in Nightingale, V. and Dwyer, T. (eds) *New media worlds: Challenges for convergence*. Melbourne: Oxford University Press, pp. 132–146.

Weatherall, J.O. (2013) *The physics of Wall Street: A brief history of predicting the unpredictable*. Boston: Houghton Mifflin Harcourt.

Webster, F. (2006) *Theories of the information society*. Third edition. London and New York: Routledge.

Weinberger, D. (2007) *Everything is miscellaneous*. New York: Times Books, Henry Holt and Company.

Weinberger, S. (2007) Son of TIA: Pentagon surveillance system is reborn in Asia. *Wired*, 22 March. Available: http://www.wired.com/politics/onlinerights/news/2007/03/SINGAPORE?currentPage=all.

Wessels, B. (2010) *Understanding the Internet: A socio-cultural perspective*. Basingstoke and New York: Palgrave Macmillan.

White, A. (2007) Understanding hypertext cognition: Developing mental models to aid users' comprehension. *First Monday*, 12(1). Available: http://firstmonday.org/htbin/cgiwrap/bin/ojs/index.php/fm/article/view/1425/1343.

White, A. (2008) The 'universal library' returns in digital form. *Libellarium*, 1(1): 111–130. Available: http://hrcak.srce.hr/index.php?show=clanak&id_clanak_jezik=58371.

White, A. (2009) A grey literature review of the UK Department for Culture, Media and Sport's creative industries economic estimates and creative economy research programme. *Cultural Trends*, 18(4): 337–343.

White, A. (2011) Digital Britain: New labour's digitisation of the UK's cultural heritage. *Cultural Trends*, 20(3–4): 317–325.

White, A. (2012) Temporal utopianism and global information networks. *The Fibreculture Journal*, 20. Available: http://twenty.fibreculturejournal.org/2012/06/19/fcj-145-temporal-utopianism-and-global-information-networks/.

Williams, M.E. (2013) What really happened to Sunil Tripathi? *Salon*, 25 April. Available: http://www.salon.com/2013/04/24/what_really_happened_to_sunil_tripathi/.

Williams, R. (1990/1975) *Television: Technology and cultural form*. London: Routledge.

WIP (2008) *The World Internet Project report 2009*.

Woodward, C. (2012) Car industry bailout was a key Obama issue. *USA Today*, 7 November. Available: http://www.usatoday.com/story/money/cars/2012/11/07/obama-romney-auto-industry-general-motors-gm-chrysler/1690071/.

Woollacott, E. (2013) Appeals Court hints strongly that Google Books project is fair use. *Forbes*, 1 July. Available: http://www.forbes.com/sites/emmawoollacott/2013/07/01/appeals-court-hints-strongly-that-google-books-project-is-fair-use/.

Work Foundation [The]/NESTA (2007) *Staying ahead: The economic performance of the UK's creative industries*. Available: http://www.theworkfoundation.com/Assets/Docs/Creative_Industries_Chapter2_3.pdf.

World Summit on the Information Society (WSIS) (2005) *WSIS outcome documents*, December 2005. Geneva: International Telecommunication Union/United Nations. Available: http://www.itu.int/wsis/outcome/booklet.pdf.

World Summit on the Information Society (WSIS) (2012) *WSIS Forum outcome document. Identifying emerging trends and a vision beyond 2015! Version 1.1*. Geneva: International Telecommunication Union/United Nations. Available: http://groups.itu.int/LinkClick.aspx?fileticket=3T8l-8df8yw%3d&tabid=2103.

World Summit on the Information Society (WSIS) Civil Society Plenary (2003) *Civil society declaration to the world summit on the information society: Shaping information societies for human needs*. Available: http://www.itu.int/wsis/docs/ geneva/civil-society-declaration.pdf.

Wyatt, S. (2008) Feminism, technology and the information society: Learning from the past, imagining the future. *Information, Communication & Society*, 11(1): 111–130.

Wynn, E. and Katz, J.E. (1997) Hyperbole over cyberspace: Self-presentation and social boundaries in Internet home pages and discourse. *The Information Society: An International Journal*, 13(4): 297–327.

Yates, F. (1966) *The art of memory*. London and Hedley: Routledge and Kegan Paul.

Younglai, R. and Spicer, J. (2009) US SEC says 'dark pools are emerging risk to market'. *Reuters*, 18 June. Available: http://www.reuters.com/article/2009/06/ 19/us-financial-regulation-sec-sb-idUSTRE55I03V20090619.

Ziegler, M. (2013) Premier League launches clampdown on pubs illegally using foreign satellites to show games. *Daily Mail/Press Association*, 24 July. Available: http://www.dailymail.co.uk/sport/football/article-2376211/Premier -League-clampdown-illegal-foreign-satellites-pubs.html.

ZoomInfo (2013) Andy White [profile]. Available: www.zoominfo.com/p/Andy -White/734713281.

Zuboff, S. (2001/1988) Dilemmas of transformation in the age of the smart machine, in Trend, D. (ed.) *Reading digital culture*. Malden, MA, Oxford Carlton, Australia: Blackwell Publishing, pp. 125–132.

Index

Note: The letters 'n' following locators refer to notes.

Printed in the United States of America